QUALITY ASSURANCE PRINCIPLES FOR ANALYTICAL LABORATORIES

FREDERICK M. GARFIELD

STOCKTON STATE COLLEGE LIBRARY
POMONA, NEW JERSEY 08240

Copyright © 1991
BY AOAC INTERNATIONAL

All rights reserved. No part of this book may be reproduced in any form or by any means without the written permission of AOAC International.

Printed in the United States of America

Printed on acid-free paper

Library of Congress Catalog Card Number 91-22409
ISBN 0-935584-46-3

First Printing 1991
Second Printing 1992

Table of Contents

QD 51 .G38 1991
Garfield, Frederick M.
Quality assurance principles for analytical laboratories

Foreword to the Second Edition i

Chapter 1   **Quality Assurance Planning** 1
Definitions 2
Establishing the Program 2
Quality Assurance Program Elements 3
The Evaluation of Costs and Benefits 4
Program Objectives 5
Quality Assurance Committee 6
The Quality Assurance Manual 7
Proposed Manual Format 7
Recommendations 10
References 11

Chapter 2   **Statistical Applications and Control Charts** . . 13
Data Presentation 13
Measures of Central Tendency 14
Measures of Dispersion 14
Normal Distribution 15
Additional Useful Applications of Statistics 17
Standard Deviation Estimated from a Pair of Results . . . 18
Combining (Pooling) Estimates of Standard
    Deviations 18
Standard Deviation Estimated from Duplicate
    Measurements 18
Comparison of Means: The t-Test 19
Confidence Interval for a Mean 19
Comparison of Two Standard Deviations: The F-Test . . 19
Important Statistical Tables 20
Outliers 20
Curve Line Fitting (Regression Line) 21
Control Charting 21
Control Limits 22
Construction of a Control Chart 22
Control Charts Based on Range 25
Guidelines on Interpretation of Control Charts 25
Cumulative Sum Charting 26
Recommendations 27
References 28

Chapter 3	**Personnel Considerations**	**30**
	The Role of the Laboratory Director	31
	The Role of Supervisors	31
	The Responsibilities of Nonsupervisory Staff	31
	Position Qualification and Position Descriptions	32
	The Preemployment Interview	33
	Orientation and Training	33
	Training Methods	35
	Performance Appraisals	36
	Self-Evaluation	37
	The Personnel File	38
	Recommendations	38
	References	39
Chapter 4	**Management of Equipment and Supplies** . . .	**40**
	Equipment Selection and Purchase	40
	Equipment Installation and Servicing	41
	Preventive Maintenance for Equipment	42
	Supply Management	44
	Chemicals	44
	Reference Standards	46
	Standard Solutions	47
	Purified Water	48
	Culture Media	49
	Volumetric Glassware	50
	Cleaning Glassware and Other Laboratory Ware	51
	Cleaning, Drying, and Sterilizing Bacteriological Glassware	51
	Recommendations	52
	References	53
Chapter 5	**Sample and Record Handling**	**54**
	Sample Accountability	54
	Maintenance of Analytical Records	58
	Retention of Records and Samples	59
	Computerized Records	59
	Recommendations	62
	References	63
Chapter 6	**Sampling and Sample Analysis**	**64**
	Sample Collection	64
	The Sampling Plan	65
	Subsampling for Analysis	68
	Sample Preparation for Analysis	70
	Method Selection	71

	Method Validation	74
	Ruggedness Testing of Methods	76
	Methods Control	76
	Accuracy and Precision	77
	Critical Control Points	78
	Analysts	79
	The Environment	80
	Instruments	80
	Mistakes (Blunders)	82
	Control Charts	83
	Corrective Actions	83
	Recommendations	84
	References	85
Chapter 7	**Proficiency and Check Samples**	**87**
	Intralaboratory Testing	87
	Interlaboratory Test Programs	89
	Proficiency Test Program Format	90
	Test Programs in the United States	93
	Recommendations	93
	References	94
Chapter 8	**Audit Procedures**	**95**
	Performance Audits	95
	System Audits	96
	The Quality Assurance Audit Unit	97
	Selection of Auditors	97
	Auditor Training	97
	Audit Planning	98
	Audit Visit	99
	Evaluation and Reporting	100
	FDA Approach to Quality Assurance	101
	Recommendations	102
	References	103
Chapter 9	**Design and Safety of Facilities**	**104**
	Laboratory Design Considerations	105
	Laboratory Safety	108
	Safety in Facility Design	110
	Safety Equipment	110
	Hazardous Materials	111
	Emergency Control Procedures	112
	Personal Habits and Safe Operating Practices	113
	Recommendations	115
	References	116

Chapter 10	**Laboratory Accreditation**	**117**
	Approaches to Accreditation	118
	Objectives of Laboratory Accreditation Systems	119
	Accreditation Criteria	119
	National Accreditation Programs	120
	Good Laboratory Practice	122
	Certification, Registration, or Licensing of Chemists . . .	123
	Conclusions	124
	References	125
Appendix A	**Typical Contents of a Quality Manual for Testing Laboratories**	**127**
Appendix B	**Forms Used by U.S. Federal Agencies**	**134**
Appendix C	**Instrument Performance Checks**	**147**
Appendix D	**FDA Audit Measure Procedures**	**158**
Appendix E	**Proficiency and Check Sample Programs** . . .	**166**
Appendix F	**Accreditation Criteria**	**174**
Index		**181**

Foreword to the Second Edition

Since first published in 1984, this handbook has had four printings and copies distributed in over 45 countries. It has been used as a management guide and as a text in the Association of Official Analytical Chemists' (AOAC) quality assurance short-course program presented in the United States, Canada, and Europe.

As was the first edition, this new edition of *Quality Assurance Principles for Analytical Laboratories* is designed to provide useful guidelines for initiating a quality assurance program, or for better organizing or improving an existing one. It provides information for laboratory directors to allow movement toward a systematic and solidly based justification for commitment of resources toward improved laboratory operations.

This publication is not an original treatise on quality assurance. It draws on published principles, practices, guidelines, and procedures used by many organizations in their programs.

New information and new concepts have developed in laboratory quality assurance since the handbook was first published. These developments suggested the need for revision. The original format has been generally retained in this Second Edition, but chapters have been rearranged with extensive changes in organization and content. A chapter has been added on statistical applications and analytical control charting. Its main purpose is to call attention to the various ways these techniques can be used in laboratory quality assurance programs.

The appendices, too, have been revised, and an important one has been added covering criteria developed by the AOAC for laboratory accreditation. These criteria can be used by laboratories for self-evaluation of their quality assurance programs and the procedures by how they manage their operations.

Forms and reliable practices, procedures, and information used by several federal agencies are illustrated in the text or in the appendices. Each chapter offers appropriate recommendations to assist the laboratory director in the development and operation of a quality assurance program.

The author is indebted to Ted Hopes for his extensive editing of and suggested modifications in the manuscript, and to Cliff Kirchmer for his review and suggestions for changes in Chapter 2, "Statistical Applications and Control Charts," and for his permission to use some of his material on the subject that he presented at AOAC short courses. The author is also grateful to the late William Cobb, who as AOAC's Editorial Board Chairman contributed so much to moving this project forward, to members of the Editorial Board, to Krystyna McIver, AOAC's Director of Publications, and to Eugene Klesta, AOAC's Quality Assurance Committee Chairman, as well as members of his committee who offered invaluable advice for revision and extension of the handbook.

Chapter 1

QUALITY ASSURANCE PLANNING

An analytical laboratory should have as one of its principal purposes the production of high-quality analytical data through the use of analytical measurements that are accurate, reliable, and adequate for the intended purpose. This objective can be accomplished in a cost effective manner under a planned and documented quality system of activities.

Growing concern with poor laboratory practices has led to a proliferation of governmental regulations relating to good laboratory practices, the initiation of laboratory accreditation programs, and the development of quality control and quality assurance courses and similar activities. These measures were precipitated in part, in the United States, by passage of the Clinical Laboratory Improvement Act of 1967 (*1*), and the U.S. Food and Drug Administration's "Good Laboratory Practices Regulations for Non-Clinical Laboratory Studies" (*2*), which became effective in June 1979. Guidelines prepared by the International Organization for Standardization, American Society for Testing and Materials, International Laboratory Accreditation Conference, and other national and international organizations, also served to stimulate the adoption of quality assurance principles in laboratory management.

It is well known that serious deficiencies can occur in laboratory operations when insufficient attention is given to the quality of the work product (*3*). Applying the necessary controls and checks to ensure this quality is not a simple undertaking. It requires not only a thorough knowledge of the laboratory's purpose and operation, but the dedication of the management staff and the operating staff to standards of excellence. Management's commitment will be demonstrated by its willingness to allocate resources and its uncompromising insistence that procedures be developed, written, and followed. This commitment may be based, in part, upon the need for the laboratory to achieve outside accreditation in order to engage in analytical contract work. If it is possible that the laboratory will be involved in litigation, then the measurements and findings must be not only scientifically credible, but legally defensible. To achieve this level of distinction, the laboratory will find it necessary to operate under a quality assurance system that includes extensive documentation of its activities.

Definitions

Some confusion may exist in our understanding of terms frequently used in laboratory quality assurance discussions. The following general terms will apply in this text; other terms will be defined later as they occur:

quality system	Laboratory activities aimed at producing accurate work and a high-quality work product.
quality control	Planned activities designed to provide a quality product.
quality assurance	Planned activities designed to ensure that the quality control activities are being properly implemented.
good laboratory practices	Official rules and operating procedures that are considered to be minimum requirements for the promotion of quality and integrity of the work product.
laboratory accreditation	Formal recognition of a laboratory by an independent science-based organization that the laboratory is competent to perform specific tests.

Establishing the Program

There is no single, generally accepted plan for establishing a laboratory quality assurance program. Each organization will have its own idiosyncrasies and problems that require special consideration and treatment. There are general principles, however, that are transferable from one situation to another with only minor modifications.

Once management has decided that it will establish a program, and is prepared to support it financially and administratively, it must develop a plan (4). This need not be complicated. First, existing laboratory operations and practices must be evaluated with respect to quality assurance needs and the QA checks and procedures currently in place. Second, quality assurance requirements will be developed, and, finally, procedures will be written that describe how those requirements will be met.

From the very beginning the entire laboratory staff must be involved in planning and developing the program, with consensus approval obtained at various steps from key affected personnel. This approach will be educational for the staff in that it will involve them in the problem-solving phase of quality assurance program development, but perhaps most importantly it will reach to the operational level to identify the critical processes, practices, and measurement approaches within the laboratory that must be considered in a well-planned program. Finally, it will provide legitimacy and authority for the actions that will be necessary to accomplish the QA objectives and will help to win commitment by the entire staff to the program that is adopted.

The responsibility for planning the quality system is generally divided among three groups: top management, the supervisory staff, and operating personnel (5). Management establishes policy, commits resources, assigns responsibilities, sets standards and policy, approves appropriate stages of the plan, and maintains general accountability. The supervisory staff assumes the responsibility for development and implementation of the program. Supervisors must anticipate processes that might affect the quality of results, seek consensus approval by employees of various segments of the plan as they are developed, obtain cooperation and involvement of the operating staff in the quality assurance effort, supervise changes of the plan, and consult with management at appropriate times. Operating personnel provide the technical expertise, technical advice and guidance, and do the actual writing and reviewing of assigned parts of the plan. They are also responsible for remaining alert to needed changes in the plan that become apparent as it is applied on a day-to-day basis. Once these responsibilities are recognized by all persons concerned, the development of the plan is ready to move forward.

Quality Assurance Program Elements

The proper and complete identification of quality elements is an important consideration in establishing a quality assurance program. These elements, in fact, encompass nearly all of the activities of a laboratory's operation, including administrative functions. The emphasis that is given to each of the identified elements will vary between laboratories depending upon the particular setting, the size of the laboratory, its purpose, and the quality of the measurements deemed necessary. The commonly recognized elements are listed below and will receive further consideration throughout the text (6, 7).

Management's policy statement	Format of the QA plan
Program objectives	Personnel practices
Cost-benefit evaluation	Work planning for quality
Control of samples and records	Chain-of-custody requirements
Procurement practices	Reference standards
Equipment maintenance	Methods selection
Methods evaluation	Intra- and interlaboratory testing
Measurement principles	Sampling (field and lab)
Safety	Statistical considerations
Laboratory Design	Audits
Reports to management	Corrective actions
Training	Program revision and update

The Evaluation of Costs and Benefits

Cost considerations are certainly important in laboratory management and in quality assurance planning. Adding a quality assurance program will increase the cost of operation, and the increased cost must be fairly judged against the benefits derived. Costs are tangible and not too difficult to assess, but most of the benefits are intangible, and evaluating their importance involves subjective judgments. The public image of the organization, the need for product improvement, the effect of government laws and regulations, and complaints from customers are examples of items that cannot be ignored. A laboratory quality assurance program must consider such matters as part of the overall plan. Other important benefits of a good quality assurance program that cannot be overlooked are improved laboratory credibility and staff morale. The savings in not having to reanalyze, correct, or even discard unreliable data, or misjudged product samples, will often justify a significant part of the increased cost of a good quality assurance program.

Management may, however, wish to develop a detailed cost-benefit analysis in order to achieve an optimum, and measurable, balance. The objective of this exercise is to minimize the cost of activities that are deemed sufficient to control data quality. The first step in this analysis is to identify the essential quality-related activities and associate them with major cost areas. The principal cost areas can be dealt with as prevention costs, appraisal costs, and correction costs (*8*).

Prevention costs are those required to keep unacceptable data from being generated in the first place. They include the costs associated with performing proper laboratory planning and documentation; preparing sound procurement specifications and criteria for acceptance of new equipment, materials, and services; providing sufficient and suitable training for laboratory personnel; following a rigorous schedule of equipment preventive maintenance; and performing the necessary system calibration to improve and maintain the accuracy of the data produced.

Appraisal costs are those required to maintain the measurement system in statistical control. Activities in this area include quality control measures that evaluate the performance of analytical equipment and procedures; independent audits of these quality control measures; interlaboratory and intralaboratory proficiency testing; and quality assurance assessment, including reporting of quality assurance activities and findings.

Correction costs are those required to correct conditions that have been found to be out of control or less than satisfactory. These include the costs for problem investigation to determine the cause of poor-quality data; the implementation of corrective and new preventive measures; and the reanalysis of samples, including in some cases resampling where invalid samples were originally taken.

Beyond these, of course, there are costs to customers that have resulted from the receipt of unacceptable data, from the need to recollect samples, or as the result of delinquent results. For private testing laboratories the loss of customer confidence, and patronage, may be the most significant consequence of the poor operational practices allowed to develop in the absence of an acceptable quality assurance program.

Some laboratories may be interested in developing a system to monitor the costs associated with quality control. Implementing such a system need be neither difficult nor expensive if existing time- and-expense reporting procedures are used. The system must be carefully planned to cover such items as scope, objectives, cost data sources to be used, evaluation of existing cost data, and reporting formats. The next step is identifying quality-related activities that are most representative of the monitoring operations. These would be essentially those identified earlier under *prevention, appraisal,* and *correction* costs. Decisions will have to be made with regard to time-and-charge rates: for example, charges for senior personnel-charges for laboratory aides.

Once the cost monitoring system is operational, quality cost analysis will be possible, enabling the laboratory director to identify areas needing attention and adjusting. These techniques of analysis can range from simple chart preparation to more sophisticated mathematical models. Two common techniques are trend analysis and Pareto analysis. Trend analysis simply compares present to past quality expenditures by category. Pareto analysis identifies the areas of greatest potential for quality improvement by listing factors that contribute to problems, ranking factors according to the magnitude of their contribution, and directing corrective actions to the largest cost contributor.

Obviously, costs are not uniformly distributed among the three categories. What is often true in medicine is true here as well, namely, a small investment in prevention can produce a large dividend, in this case in saved appraisal and correction costs.

It has been estimated that the cost increase of a quality assurance program is in the range of 10–20% of operational costs. Most of this increase is assignable to control sample analyses, the use of approved specifications and standard materials, and equipment preventive maintenance.

Program Objectives

The quality assurance program requires a set of precise objectives. These must be clearly stated and understood by both management and staff. If there is a good understanding of the objectives, the program will function more efficiently and there will be less need for day-to-day problem solving. Objectives should relate to the results desired, not to specific processes or activities. Furthermore, objectives provide a yardstick against which the success of the program can be measured. The following are just a few of the more common objectives of a quality assurance system (*9, 10*):

- To upgrade the overall quality of laboratory performance
- To maintain a continuing assessment of the quality of data generated by analysts
- To identify good analytical methods and research needs
- To provide a permanent record of instrument performance as a basis for validating data and projecting repairs and replacement needs

- To ensure sample integrity
- To improve record keeping
- To produce analytical results that can withstand legal scrutiny
- To detect training needs

We may add to this several quite general objectives, such as the improvement of productivity, establishment of the laboratory's credibility and reputation to satisfy the requirements of clients or users of the laboratory's services, and fulfilling the requirements for accreditation.

Quality Assurance Committee

Many laboratories have found it beneficial to establish a quality assurance committee to help in the development of the quality assurance program. These committees are typically composed of representatives from the supervisory and scientific staffs, with the quality assurance coordinator of the facility designated as the chairperson of the committee. Such a committee works well for several reasons. It seems to reduce staff resistance and negativism toward the concept of a mandated program, and it allows individual members to take responsibility for writing sections of the program. This participation is instrumental in gaining the understanding and cooperation of the staff. Once the basic program has been developed and a manual prepared, the committee can be disbanded, or even retained for a time to monitor revisions of the manual.

As a general rule, the chairperson of the committee is the person to whom quality assurance coordinator duties are assigned, as mentioned above. It is preferable that the position of quality assurance coordinator be located organizationally outside of the laboratory, and not be involved in normal day-to-day laboratory activities. The coordinator should be responsible to management. In some laboratories this arrangement may not be possible because of the small size of the staff, but it is essential that the person selected for these duties, even though a part of the laboratory operation, be one who can be objective about the duties and be capable of coordinating all of the quality assurance activities, retaining the confidence of both management and staff.

The responsibilities assigned to the coordinator can be broad or limited as management sees fit. The coordinator may or may not have the authority to initiate changes in the program — most often this authority remains with the laboratory director. Duties for the coordinator can include all or several of the following:

- Monitoring quality assurance activities to determine conformance with policy, procedures, and sound practices
- Making recommendations for corrections and improvements in the program
- Seeking out and evaluating new ideas and developments in quality assurance and, when indicated, recommending their adoption
- Evaluating control charts, instrument preventive maintenance records, and other reports generated under the quality assurance program

- Advising management with regard to changing technology, new analytical methods, and new analytical equipment
- Coordinating or conducting quality problem investigations
- Participating in laboratory audits
- Recommending staff training to improve the quality of operations

In large, multidiscipline laboratories it may be necessary to establish a quality assurance *unit* rather than an ad hoc committee. Such a unit must be independent of the laboratory, conduct full quality assurance audits, and report to a higher level of management. The organization, training, and responsibilities of such a unit are discussed in Chapter 8, "Audit Procedures."

The Quality Assurance Manual

The importance of written procedures when a precise level of performance is required cannot be overemphasized. It is important for purposes of clear understanding and periodic reference that each laboratory have its quality assurance procedures incorporated into a set of directives, usually in a manual. This manual is the written document that identifies the policies, organizational objectives, functional activities, and specific quality activities designed to achieve the quality goals desired for the operation of the system (*11*). The manual describes, in a reasonably systematic way, the measures the laboratory will employ to implement the quality assurance program.

The manual needs to be flexible and adaptable to changes in methods, techniques, and personnel. Regardless of how well written the program is, it will be circumvented, or ignored altogether, if it tends to stifle programs or is allowed to stagnate. Measures and procedures described in the manual require constant attention and unconditional adherence in day-to-day application. As with any serious rule (or law) that governs behavior, there is a proper way to change it, but until it *is* changed it must be respected and treated as *the rule*.

Presently there is no national or international standard format for such a manual. This is understandable because the level of quality requirements is quite different among laboratories, largely because of differences in their size, fields of activity, type of work, and organizational complexity. Depending on its overall purpose, the manual can be relatively simple in structure or quite complex and detailed. A reasonable format is shown below, followed by examples of those used by others.

Proposed Manual Format

The quality assurance manual should contain the following elements (*12*):
- Title page, policy statement, management approval signatures, and table of contents
- Organization of the laboratory and description of its responsibilities, including an organizational chart showing key positions
- Quality assurance objectives with respect to accuracy, precision, completeness, time frames, and so forth

- Description of the quality assurance procedures and requirements
- Performance audits describing specific procedures for the planned audits to be used to assess the success of the quality assurance program
- Short-term and long-term corrective actions that must be taken to address specific problems
- Instructions for the use of forms and the type and format of reports that must be made to management, including the frequency of those reports
- Description of the proper mechanism for updating and correcting the manual and the procedures described in it, and of distribution of the manual and its updates

The matter of quality assurance manual format has been given extensive attention by the International Laboratory Accreditation Conference (ILAC), a group comprising representatives from about 50 countries interested in ways to obtain accreditation recognition among laboratories in these countries. This organization is given additional attention in Chapter 10, "Laboratory Accreditation."

A special task force of ILAC was assigned the problem of developing a format for a quality assurance manual, and in 1984 produced a document titled "Typical Contents of a Quality Manual for Testing Laboratories" (13). According to the task force report, "The document constitutes guidelines intended to assist a laboratory which proposes to set forth, in a reasonably systematic way, the measures it employs to implement its internal quality system. It is intended as guidance to the laboratory in development of its own quality system, and in the preparation of a Quality Manual that describes the elements and functioning of the system."

Because of the length of the ILAC report, the guidelines are paraphrased below, and the major recommendations, taken from the report, are reproduced in Appendix A. As a minimum, the summary and the recommendations can serve as checklists of items to be considered in producing a quality assurance manual.

- **QA POLICY:** Describing the objectives of the program, the commitment by management to use resources to implement it, and a general statement on how it will be managed.
- **DESCRIPTION OF THE MANUAL:** How it will be managed (for example, amended, supplemented, and distributed), definitions of unusual or ambiguous terms, and the areas of laboratory work the manual is to cover.
- **LABORATORY ORGANIZATION:** The responsibilities of management and operating personnel for quality assurance.
- **EQUIPMENT:** Procedures for inventorying, maintaining, and testing laboratory equipment.
- **ENVIRONMENT:** Control specifications and monitoring procedures.
- **ANALYTICAL METHODS:** Criteria for the selection of testing methods and procedures for method validation.
- **SAMPLE HANDLING:** Disbursement, storage, disposal, and special security procedures, including record keeping.
- **ANALYTICAL RESULTS:** Requirements for reporting, verifying, and validating analytical results.

- **RECORDS:** Maintenance of records and the requirements for confidentiality.
- **AUDIT PROCEDURES:** Evaluating the effectiveness of the QA program, proficiency testing requirements, and taking corrective actions.
- **CONTRACT WORK:** Procedures for verifying the correctness of data received from external contract facilities.

Dux (14) suggests that the best format for the quality assurance manual is a series of standard operating procedures (SOPs). These SOPs are instructions on how to perform tasks and descriptions of the approved or required procedures for accomplishing the quality assurance objectives. He suggests that each SOP take the following form:

- Number — for reference purposes
- Title — A brief but descriptive title so that the SOP can be easily identified in an index
- Background information (optional)
- Scope or field of application
- Purpose of the SOP — if a good reason for the SOP cannot be described, according to Dux, the SOP should not be written
- Operations — numbered paragraphs to take the user through the actions that would achieve the purpose of the SOP

There are several other documents that have been developed that can be helpful to laboratories bent on developing a quality assurance program (15–18). Especially worthy of review is a guideline prepared by the Standards Council of Canada that contains detailed information on the preparation of a manual (19).

A formal approach to the preparation of the various sections of the manual can help promote the quality of the manual itself. The purpose of written procedures is to facilitate communication, so the people selected to write the manual should not only understand the technical requirements of the assignment, but have a suitable level of writing skill. Each completed unit should be subjected to critical review to resolve questions, remove ambiguities, and produce a document that the concerned parties can agree upon. Management should issue the document with the appropriate authorizing signatures.

Regular review of the manual and the program must be planned from the beginning. A new program may require more intense and frequent scrutiny during its first months in use, but the time between review periods can be lengthened once the program becomes established. The program must remain flexible and adaptable to changes that are brought about by experience, new technologies, new developments in quality assurance, new organizational purposes and needs, and changing personnel situations. Similarly, the program must not be allowed to stagnate — only strict enforcement of the program as written, or revised, will prevent it from becoming a worthless document.

Recommendations

Before embarking upon the development of a quality assurance program, or the modification or improvement of an existing one, do the following:
1. Develop a plan and include supervisors and staff in the planning operation.
2. Consider the cost-benefit equation, and especially take into account the time that the laboratory is willing to devote to program development and enforcement.
3. Set objectives, giving due consideration to the level of the quality of work needed or desired.
4. Consider the quality assurance elements appropriate to the operation, giving thought to the various facts discussed in the chapters that follow.
5. Appoint a quality assurance committee or unit, and assign a quality assurance coordinator as chairperson. Clarify their responsibilities.
6. Decide on the format of a quality assurance manual and assign the writing of various sections to responsible individuals.
7. Have the manual sections reviewed by supervisors and selected key scientific personnel, and make changes as necessary after group discussions.
8. Implement the program, monitor it, and make changes as need and experience dictate.
9. Document all activities.

References

(1) "Clinical Laboratory Improvement Act of 1967," Part F, Title III, Public Service Health Act, Section 353

(2) *Fed. Regist.* (Dec. 22, 1978) "Non-Clinical Laboratory Studies, Good Laboratory Practice Regulations," p. 60013 (21 CFR Part 58)

(3) Horwitz, W. (1977) "Good Laboratory Practices in Analytical Chemistry," presented at the 17th Eastern Analytical Symposium on "Good Laboratory Practices," New York, NY

(4) Taylor, J.K. (1987) *Quality Assurance of Chemical Measurements,* Lewis Publishers, Inc., Chelsea, MI, p. 234

(5) Freeberg, F.E. (1980) in *Optimizing Chemical Laboratory Performance Through the Application of Quality Assurance Principles,* F.M. Garfield et al. (Eds), Association of Official Analytical Chemists, Arlington, VA, p. 15

(6) Ratliff, T.A. (1980) in *Testing Laboratory Performance: Evaluation and Accreditation,* NBS Publication 591, G.A. Berman (Ed.), National Institute of Standards and Technology, Gaithersburg, MD, p. 104

(7) "NIOSH Specifications for Industrial Hygiene Laboratory Quality Program Requirements" (1976) National Institute for Occupational Safety and Health, Cincinnati, OH

(8) Wilcox, K.R., et al. (1977) in *Quality Assurance Practices for Health Laboratories,* S.L. Inhorn (Ed.), American Public Health Association, Washington, DC, p. 20

(9) *Instrumentation Quality Assurance Manual* (1979) U.S. Consumer Product Safety Commission, Washington, DC

(10) "Baltimore District Intra-laboratory Quality Assurance Program" (1984) U.S. Food and Drug Administration, Baltimore, MD, Memo, March 25, 1983

(11) "Guidelines and Specifications for Preparing Quality Assurance Project Plans" (1980) U.S. Environmental Protection Agency, Municipal Environmental Research Laboratory, Cincinnati, OH

(12) *Quality Assurance Handbook for Air Pollution Measurements,* Vol. 1 (1976) U.S. Environmental Protection Agency, Research Triangle Park, NC, Sec. 1.4.23

(13) Report of Task Force D (1984) "Typical Contents of a Quality Manual for Testing Laboratories," International Laboratory Accreditation Conference, Secretariat: Standards and Quality Policy Unit, Department of Trade and Industry, London, England

(14) Dux, J.P. (1986) *Handbook of Quality Assurance for the Analytical Chemistry Laboratory,* Van Nostrand Reinhold Co., New York, NY, p. 95

(15) "International Standards Organization, Guide 25" (1982) American National Standards Institute, New York, NY

(16) "Standard Guide for Establishing a Quality Assurance Program for Analytical Laboratories" (1983) American Society for Testing and Materials, Philadelphia, PA

(17) "Facteurs de la Qualite (Quality Factors)" (1983) Reseau National D'Assais, Paris, France

(18) "A Quality System Standard for Laboratories" (1982) American National Standards Institute, New York, NY

(19) "Guidelines for Preparing a Quality Manual for Testing Organizations" (1987) Standards Council of Canada, Ottawa, Ontario, Canada

Chapter 2

STATISTICAL APPLICATIONS AND CONTROL CHARTS

One statistician has remarked that statistics is a powerful management tool in quality control that, unlike most other tools, becomes sharper with use (*1*). Another suggests that statistical methods are not an optional extra, but an integral part of the processing and interpretation of numerical evidence for decision and action. There is no choice whether or not to use statistics, only how well they should be used (*2*).

This chapter contains a brief description of basic statistics and control charting procedures used in quality control. It is not intended as a substitute for discussions of these subjects that provide more rigorous and detailed treatment (*3*). The main purpose of this chapter is to call attention to various ways in which these statistical techniques can be helpful in the quality assurance program.

Analytical control charting, also discussed in this chapter, is based upon principles developed by W.A. Shewhart in 1939. It provides a means, using a choice of control limits, for distinguishing between a pattern of random variations and one that arises from assignable causes of variation. It is a tool that can provide meaningful and early information that can be used to control a process.

Data Presentation

Before any statistical calculations can be performed on data, where many observations are recorded, it is desirable and necessary to arrange and condense the data so that the main features of the assembly are clear. This can be done in a qualitative way by grouping the measurements, by forming frequency tables and charts, or by using descriptive statistics. The data to which statistical methods can be applied may be measurements made on individual elements or counts of the number of elements that possess specific attributes (*4*).

It is difficult to evaluate the information contained in a large set of measurements if data are shown in the order in which they are received. Data in this form is referred to as *raw data*. An example of raw data would be:

| 1.65 | 1.15 | 1.00 | 1.05 | 1.35 | 1.65 | 1.15 | 1.35 | 1.15 | 1.67 |

Data can be arranged in order of magnitude, in which case it is referred to as *ranked data* or *sorted data*. In this form, simple inspection will disclose such features as steady progression, a hump, a gap, or a biased frequency distribution. The same data as above, ranked, is as follows:

| 1.00 | 1.05 | 1.15 | 1.15 | 1.15 | 1.35 | 1.35 | 1.65 | 1.65 | 1.67 |

The data can be grouped into intervals and the number of values that fall within each interval counted. The same data may be presented graphically as a bar chart or histogram, where the height of each bar represents the frequency of the values in the interval. These and other techniques can be used to succinctly summarize the data and to show its characteristic patterns. It is important to recognize and to remember that only science can generate the data; statistics is the art of *presenting* the data.

Measures of Central Tendency

It is often useful to calculate a single value that will represent a body of data. Such a value is referred to as a *measure of central tendency,* and sometimes as "location parameter." The arithmetic mean, the median, and the mode are such measures, the most commonly used being the *arithmetic mean* or average ($\bar{x}$). It is the sum of the individual values in a set divided by the number of values:

$$\bar{x} = \frac{\Sigma_n x}{n} = \frac{x_1 + x_2 \ldots + x_{10}}{n} = \frac{13.17}{10} = 1.317$$

The *median* is the middle value of a ranked set of data and is often a more meaningful location parameter than the mean. If there is an even number of values in a set, the median is the arithmetic mean of the two middle values. For example, for the above set of 10 values, the two middle values are 1.15 and 1.35, and the median is 1.25, as shown below:

$$Median = \frac{x_5 + x_6}{2} = \frac{1.15 + 1.35}{2} = 1.250$$

The *mode* is the value, or values, occurring most frequently in a sample of data. The mode of the set shown above is 1.15, a value that occurs three times.

Measures of Dispersion

In addition to knowing a location parameter that is representative of a set of data, it is also important to know the degree of scatter, or dispersion, of the individual values. Generally, measures of both central tendency and dispersion are necessary to describe a set of data. The most widely used measures of dispersion are the *range,* the *standard deviation* or *relative standard deviation,* and the *variance*.

The *range* is the difference between the minimum and the maximum values of a set of measurements. Since the range makes use of only two values out of a set, it is significantly affected by the extreme values, including any "outliers" at either end of the range. For the set shown above, the range is

$$x_{10} - x_1 = 1.67 - 1.00 = 0.67$$

Mathematically, the *sample variance* is the sum of the squares of the differences between the individual values of a set and the arithmetic mean of the set, divided by one less than the number of values. The divisor $(n - 1)$ is used, rather than n, so that the value of s^2, the sample variance, is an unbiased estimate of the population variance (in the long run the average of the estimates of the sample variance will be equal to the population variance).

$$s^2 = \frac{\Sigma(x-\bar{x})^2}{n-1} = \frac{(1.00-1.317)^2 + \ldots + (1.67-1.317)^2}{10-1} = \frac{0.6401}{9} = 0.067112$$

For computational purposes, to minimize round-off errors, the equation for s^2 can also be written as:

$$s^2 = \frac{\Sigma x^2 - \frac{(\Sigma x)^2}{n}}{n-1} = \frac{17.9489 - \frac{(1.317)^2}{10}}{9} = \frac{0.60401}{9} = 0.067112$$

The *standard deviation* is the square root of the sample variance. The property of the standard deviation that makes it most practically meaningful is that it is in the same units as the observed variable x.

$$s = \sqrt{s^2} = \sqrt{0.067112} = 0.259$$

The *coefficient of variation*, preferably referred to as the *relative standard deviation*, is the ratio of the standard deviation to the mean, multiplied by 100 to convert it to a percentage of the average. In many practical problems the RSD (CV) is essentially constant over the range of interest and is a most useful measure of variation.

$$\text{RSD(CV)} = \frac{s}{\bar{x}} \times 100 = \frac{0.259}{1.317} \times 100 = 19.7$$

Normal Distribution

The most widely used continuous frequency distribution is the "Gaussian" or normal curve (Figure 1) (5). It is described by the mean (μ) and the standard deviation (σ).

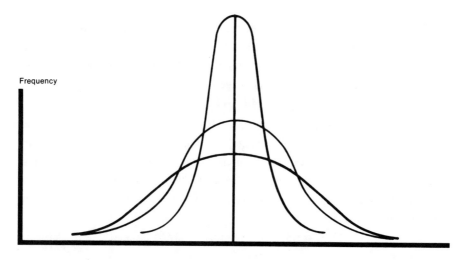

Figure 1 Normal Curve with Different σ Values

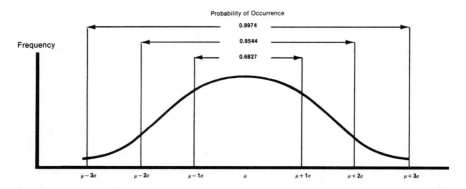

Figure 2 Probability of Occurrence Between Specified Intervals of Standard Deviation Units

As the standard deviation increases, the curve becomes more spread out and the frequency density around the mean decreases. Conversely, when the standard deviation decreases, the curve becomes narrower and the frequency density around the mean increases. When the mean changes, the curve shifts to the right or the left on the horizontal axis. The area under the normal curve between two ordinates expresses the probability that a random, unbiased measurement from a normal population will fall in the interval bounded by the two points. Probabilities for selected intervals, in units of standard deviation, are shown in Figure 2.

For example, if the distribution is "normal," the range represented by the value of the mean ± 1 standard deviation will encompass 68.27% of all readings taken; the value of the mean ± 2 standard deviations will encompass 95.44% of all readings taken; and the value of the mean ± *three* standard deviations will encompass 99.74% of all readings taken. The probability of a reading exceeding three standard deviations when a process is in control (that is, follows a consistent persistent pattern that we assume approximates a normal curve) is small, namely the difference between 100% and 99.74%, or 0.26%. These calculations have practical value for describing what should be expected, and what can be considered to indicate an out-of-control situation. Examples will be presented later in the chapter when control charts are discussed.

Additional Useful Applications of Statistics

Analytical results are important in the decision-making process. Most uses of quantitative analytical results require their comparison with a fixed value, such as a standard, or with other analytical results. The purpose of an analytical quality control program, and the use of statistics within that program, is to identify and control errors (6).

The distinction between a *population* and a *statistical sample* needs to be understood (7, 8). A *population,* or universe, is a theoretical entity defined as an entire group of people, things, or events that have at least one trait in common. In this book, a population is a collection of all *possible* analytical results from a given lot of material using a given chemical method. A population can be characterized using descriptive values called "parameters," represented by lower-case Greek letters. Two of these are μ (mu), which represents the population mean, and σ (sigma), which represents the population standard deviation. These values, too, are theoretical since they do not represent, and are not based upon, any measurements of the physical world. They can be conceived of as all the possible analytical measurements of the analyte of interest.

A *sample* is represented by a set of data obtained from a population; it is a finite number of cases, instances, measurements, or observations taken of things that qualify as members of the population. The information that we *collect* from the sample units is used to estimate population parameters. Thus we make a small set of analytical observations on a sample from a large lot of goods to evaluate the quality of the lot. Descriptive values that are applied to samples are represented by Italic letters, for example $\bar{x}$ represents the sample mean, and *s* represents the sample standard deviation.

It is important to understand that the mean of a number of observations is a more reliable estimate of the population mean than is a single observation (9). The standard deviation of a series of mean values, therefore, would be much smaller than the standard deviation of a series of individual measurements, since the mean values would tend to approximate more closely the one "true" value of the population, whereas individual measurements would be subject to many sources of random variability. This standard deviation of sample means is usually known as its *standard error* and is often referred to as the standard error of the mean (10). This value expresses the amount of variability among the various sample means. It can be calculated from the standard deviation *s* and the number of observations *n*:

$$s_{\bar{x}} = \frac{s}{\sqrt{n}}$$

In addition to the standard deviation of the mean being smaller (by a factor of $1/\sqrt{n}$) than the standard deviation associated with observations or results, the distribution of the sample means tends rapidly toward the normal as the number of means available increases. This is known as the *central limit theorem*. The theorem states that when successively larger and larger random samples are taken from a population, the distribution of the means of these samples assumes the shape of the normal curve, regardless of the distribution of individual measurements.

Standard Deviation Estimated from a Pair of Results

In the case of a single duplicate pair of results, the equation for the standard deviation can be reduced to the following, where D is the absolute difference between the members of the pair (6):

$$s = \frac{D}{\sqrt{2}}$$

Combining (Pooling) Estimates of Standard Deviations

It is possible to pool (combine) several estimates of standard deviation to obtain a better estimate of standard deviation (i.e., with more degrees of freedom[1]) using the equation below, where v_i is the number of degrees of freedom of the ith estimate of s.

$$s_p = \frac{\sqrt{v_i \, s_i^2}}{\Sigma v_i}$$

Standard Deviation Estimated from Duplicate Measurements

Estimates of standard deviation can be made from duplicate measurements after determining the differences of several sets of duplicate determinations. The following equation is used in which K is the number of sets of duplicates, and D is the difference between duplicate determinations. This equation is derived from the general pooling equation and the equation for standard deviation of a single duplicate pair.

$$s = \frac{\sqrt{\Sigma D^2}}{2K}$$

---

[1] Degrees of freedom is the number of independent values used to make an unbiased estimate, for example, estimating a population average using n values uses n degrees of freedom; estimating the variance of these values uses $n-1$ degrees of freedom.

Comparison of Means: The *t*-Test

The *t*-test can be used to determine whether or not there is a statistically significant difference between the means of two sets of values at a given probability level. The statistic *t* is given by the following equation:

$$t = \frac{\bar{x} - \bar{y}}{s_m}$$

The two means are $\bar{x}$ and $\bar{y}$, and s_m is the pooled standard deviation. Without going into detail, s_m is calculated in different ways, depending on whether the standard deviations for both sets of results are equal, or the standard deviations for the two sets of results are unequal. The value obtained for *t* is then compared with a tabulated value of *t* that corresponds to the chosen α, the level of significance. This comparison permits two alternatives: the first decides that there is a real difference between the two averages; the second decides that there is insufficient evidence to justify a claim that the samples differ. Additional information on the *t*-test and these calculations can be found in reference (3) of this chapter under Dewey and Wilson.

This type of testing procedure, as exemplified by the *t*-test procedure, is known as significance testing and it involves the formulation of the "null-hypothesis." In the example cited above, the equation is used for comparison of two sample means.

The *t*-test can also be used to test whether a sample mean is significantly different from an expected or reference value *E*. In the equation below, *s* is the estimated standard deviation of a single result, and *n* results have been used in calculating the mean. The degrees of freedom for *t* are *n*–1.

$$t = \frac{\bar{x} - E}{\frac{s}{\sqrt{n}}}$$

Confidence Interval for a Mean

Confidence intervals or confidence limits are used to denote limits around a sample mean, $\bar{x}$, that encompasses the true mean of the population, μ, a given percentage of the time. This is a frequently used statistical calculation, and the confidence values are generally 90%, 95%, or 99% (in terms of α: .10, .05, and .01, respectively). In the following equation the value for *t* will depend upon the confidence desired and the degrees of freedom. The value can be found in a table identified as Student's *t* variate or distribution.

$$\bar{x} \pm \frac{ts}{\sqrt{n}}$$

Comparison of Two Standard Deviations: The *F*-Test

One statistical test for significance was developed by Sir Ronald Fisher, and is therefore called the *F*-test. The test is devoted to evaluating data to determine whether one population is more variable than another. It can be used to determine whether or not two methods differ in their precision, or whether two analyst or instrument or laboratory produces more precise results than another. In the comparison of 2 standard deviations, the statistic F is calculated as follows:

$$F = \frac{s_1^2}{s_2^2}$$

where s_1 is the higher of the two standard deviation estimates, based upon n_1 measurements, and s_2 is the lower of the two standard deviation estimates, based upon n_2 measurements.

The observed value of F is compared with F_α (df_1, df_2) which is obtained from Variance Ratio Tables ("F-Tables") that are based on the respective degrees of freedom for the estimates of s_1 and s_2. If F is greater than F_α, then s_1 can be considered to be greater than s_2 at the chosen level of confidence. If F is less than F_α, then there is no reason to believe that s_1 is greater than s_2 at the chosen level of confidence. If there is no previous knowledge of which of the two populations has the higher standard deviation, the test is performed using $F_{\alpha/2}$ (df_1, df_2) instead of F_α (df_1, df_2).

Important Statistical Tables

A variety of statistical tables can be found in the appendices of the statistical texts cited in the "References" section of this chapter. These tables are used when measurement data are being statistically analyzed. None of the tables is reproduced in this handbook since they are so readily available elsewhere. Here are the names of a few of these tables that will be important to statistical evaluation of data:
- Percent area under the normal curve between $\bar{x}$ and z
- Critical value for t-level significance for 1 and 2 tail tests
- Critical values of F for the analysis of variance
- Critical values of the chi square
- Student's t variate or distribution
- Critical values in the Dixon test for outliers
- Critical values for the Grubbs test for outliers
- Table of random numbers
- Percentage points of the t-distribution

Outliers

In a set of observations one can encounter an unusually large or small value or measurement that does not belong with the other values or measurements. This type of value is usually referred to as an outlier. There are many possible reasons for an outlier, such as blunder, malfunction of the method, calculation error, error in transcribing data, or some unusual loss or contamination. When a result appears to be an outlier, it can be subjected to a test to see whether it can reasonably be rejected. The usual purpose of the test is to identify the need for closer control of the process, or to eliminate the outlier prior to evaluation of the data. This data preprocessing should be done with caution, since it is not advisable to presume that test data are defective without first trying to identify the cause of the problem. If an outlier is discarded, that fact should be reported.

There are several statistical tests in common use for identifying outliers for normal populations. Two such tests are the Grubbs Test and the Dixon Test. These tests are not difficult to follow, but they will not be described in this text. Consult any of the statistical texts listed in reference (3) of this chapter such as Taylor (1987), Wernimont (1985), or Youden and Steiner (1975).

Curve Line Fitting (Regression Line)

Fitting a straight line to a plot of data, for example in constructing a calibration curve, is a frequent and important task for the analyst (11). A single straight line that lies closest to all of the points on a scatter plot is known as a regression line. The line can be used for making certain predictions when the following values are known: the slope of the line and the point of intercept (intersection) with the y-axis.

The equation for a straight line is $y = a + bx$. It is found from data $y_1, y_2, \ldots y_n$ associated respectively with $x_1, x_2, \ldots x_n$. If only the observations of y are affected by significant error, the least-squares-estimates for the parameters for the linear model are:

$$a = \frac{(\Sigma x_i^2 \Sigma y_i - \Sigma x_i \Sigma x_i y_i)}{c}$$

$$b = \frac{(n\Sigma x_i y_i - \Sigma x_i \Sigma y_i)}{c}$$

where

$$c = n\Sigma x_i^2 - (\Sigma x_i)^2$$

The estimate of the variance of a single observation of y is then given by

$$s^2 = \frac{[\Sigma(y_i - Y_i)^2]}{n - 2}$$

where Y_i are the values (corresponding to x_i) computed from the equation $y = a + bx$. The sum of the deviations squared will be a minimum, that is, any other straight line that can be drawn would have a greater sum of squared deviations.

Additional information on this subject can be found in reference (3) in this chapter under Eskschlager (1969).

Control Charting

A useful tool in the quality assurance program is control charting. It is best used for routine analyses that are prone to error (usually increased bias, but sometimes increased variability). A control chart is a graphical plot of test results with respect to time or sequence of measurements, with limits drawn within which results are expected to lie when the analytical scheme is in a state of "statistical control." A procedure is in statistical control when results consistently fall within established control limits. Wernimont (12) suggested that a process is in this condition when significant assignable causes of variation have been corrected or removed, allowing a set of measurements to be used to predict limits of variation and to assign a level of confidence that future measurements will be within these limits. In addition to identifying results that are out of control, a chart will disclose trends and cycles, and thus is a tool to provide real-time analysis of data and information upon which appropriate corrective action can be based.

Control Limits

According to the American Society for Quality Control (ASQC) (13), control limits are the limits on a control chart that are used as criteria for action or for judging whether a set of data does or does not indicate lack of control. There are two types of control limits generally used: warning limits, and action limits. The warning limits correspond to ±2 standard deviations from the mean, whereas action limits are set at ±3 standard deviations from the mean. One can control around the mean or around the true value. In the ideal case, where unbiased methods are being used, one would control around the true value. This would apply, for example, to precision control charts for standard solutions. For other situations there may not be a true or reference value.

There is only a 5% chance that a result will exceed the warning control limits based solely on random-error considerations, and only a 0.3% chance that a result will exceed the action limit. Exceeding the control limits indicates that precision has worsened or that systematic error may be present. The greater the deviation from the mean, the more likely that systematic error is present.

Construction of a Control Chart

Shewhart control charts are the easiest to construct, use, and interpret. Probably their most useful characteristic is the identification of individual outlying results. To set up a control chart, measurements must be gathered when the process is in control (14). This means that the analyst must be familiar with the method, must have explored the various sources of error (for example, sampling procedures, instrumental techniques, and analyte separation) and must have performed sufficient method validation to be certain that the results are acceptable. The following are reasonable steps to be followed in constructing a control chart (15):

- Identify the essential elements that compose the chart
- Decide on the goal
- Select the characteristics to be measured
- Decide how, where, and when the characteristics are to be measured
- Decide what set size is appropriate
- Define the procedure for obtaining random samples
- Obtain reliable estimates of the process mean (if not otherwise known) and of the long-term standard deviation

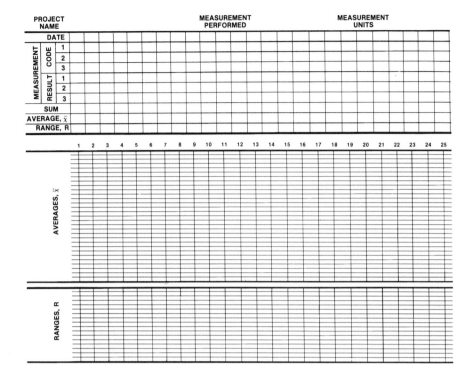

Figure 3 Shewhart Control Chart Form

❑ Lay out the center line, and the warning and action limits

Figure 3 is a standard form for a variables control chart ($\bar{x}$ and R). Figure 4 is an example of a chart based upon data obtained over 17 days for $\bar{x}$ and R (5). The use of the range, R, is discussed later in this chapter.

Kirchmer (6) suggests the use of several types of control charts and recommends an order of priority for the implementation of analytical control:

1. Precision control charts based upon standard solutions.

    The simplest type of chart is that based on standard solutions. Generally, an analysis of a control standard should be run once for every 20 analyses. If fewer analyses are performed, then one control standard should be examined along with that run. For best results, about 20 determinations should be made before calculating the statistics.

    Usually a laboratory analyzes standards at a single concentration; however, if the laboratory director wishes to present standards as unknowns, the concentration can be varied from 80 to 120% of the concentration used to establish the initial estimate of standard deviation. In this narrow range, the standard deviation does not vary significantly.

2. Precision control charts based upon duplicate analyses.

    The test samples examined for these charts should be within a narrow concentration range. If the concentrations vary over a wide range, it may be necessary to construct a number of charts each having a relatively narrow

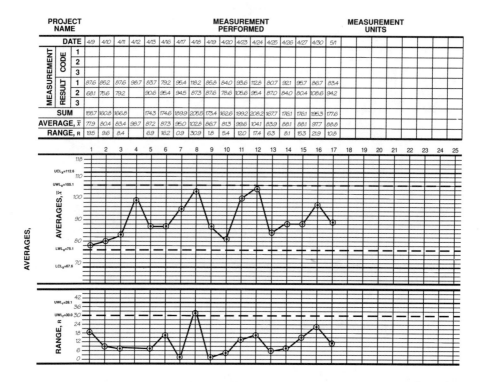

Figure 4 A Completed Shewhart Control Chart

range. Standard deviation usually depends upon concentration and the standard deviation of differences between duplicate analyses also depends upon concentration. The control chart is constructed by analyzing at least 10 duplicate test samples (20 analyses). The differences between the paired results are plotted with zero as the expected value. The second result is subtracted from the first and the differences are plotted with due regard for their sign. The standard deviation, SD, is calculated using the following equation, where N is the number of test samples that have been analyzed in duplicate:

$$SD = \frac{\Sigma d^2 - \frac{(\Sigma d)^2}{N}}{N-1}$$

3. Bias control charts based upon "spiking" recovery tests.

These control charts are intended to determine bias in analyses and provide evidence of the lack of statistical control of appropriate parts of the measurement process. Spike recovery tests may fail to find some interfering substance effects that are independent of the concentration of the analyte. Spike recovery tests should be applied only in cases where the relative standard deviation is known to be small and the equivalent concentration of the spike added lies between 1 and 10 times the concentration of the analyte in the unspiked sample. According to Kirchmer (6), spike recovery control charts are of limited value except when (a) the within-batch standard deviation, σ_w, is independent of concentration; (b) recovery tests can be applied so their analyte concentrations are the same; and (c) separate control charts are kept for each of a number of narrow ranges of concentration in the unspiked matrices.

4. Plots of values of blank determinations on a segmented chart.

These are not true control charts because there are no control limits on values. The chart is simply a plot of the responses sequentially obtained, with zero as the baseline value. Blank measurements exhibiting unusually high values stand out and provide warnings, especially of contamination or reagent deterioration.

Control Charts Based on Range

Control charts for precision can be prepared using range in place of standard deviation (6). These are categorized as R-charts, where R signifies the difference between duplicate analyses on the same test sample. This approach is similar to the one in which the standard deviation of the differences were plotted. In this instance, the range is used as a measure of precision. The following procedure is suggested by Kirchmer (6):

1. The average range ($\bar{R}$) is calculated from the equation $\bar{R} = \Sigma R/n$ based on 10–15 duplicate analyses at different times.

2. The Upper Control Limit (UCL$_R$) is calculated from UCL$_R = D_4\bar{R} = 3.27\bar{R}$. The value for D_4 is taken from a table of "Factors for Computing Control Chart Lines Using R," selected according to the number of observations (n) used to determine the range. For duplicates, $n = 2$ and the D_4 factor is 3.7. For $n = 3$, the factor is 2.58, and for $n = 4$ the factor is 2.28. The UCL$_R$ corresponds to the 99% confidence limit.

The Upper Warning Limit is calculated from UWL$_R = \frac{2}{3}(D_4\bar{R} - \bar{R}) + \bar{R}$. For duplicate samples this reduces to UWL$_R = 2.51\bar{R}$. The UWL$_R$ corresponds to the 95% confidence limit. It is suggested that at least one duplicate be run for each 10 analyses.

An important advantage of the R-chart is that no special check sample is needed. It is especially useful where standard materials are difficult to obtain; however, the duplicate determinations must be completely independent, that is, not performed simultaneously.

Guidelines on Interpretation of Control Charts

There are a number of commonly used indicators for interpreting x-bar charts to indicate when the measurement process is out of control (14): One or more points outside the three-standard-deviation limit; two or more consecutive points outside the two-standard-deviation limit; or a series of seven or more points above or below the mean, or an increasing or decreasing trend. It is often better to use a combination of these indicators rather than a single one. Interpretation of the R-chart is somewhat similar to the $\bar{x}$-chart. Five or more consecutive points above the 50% confidence limit indicates a tendency toward the process being out of control. More than one point above the UWL_R in 20 duplicate determinations is cause for concern. Any point above the UCL_R should be cause for alarm. Westgard et al. (16) discuss the performance characteristics of these indicators. It is desirable to make the process as sensitive as possible in order to detect situations that are out of control, taking care that data that are really in control are not rejected. When a decision is reached that a process is out of control, the chart is marked, analyses are stopped, and the cause or causes are investigated. When correction is made, and the system is again in control, charting can be resumed.

There are many causes for either a shift or trend in the charts: for example, incorrect preparation of a standard or a reagent, sample contamination, incorrect instrument calibration, poor analytical technique, and deviation from the method.

Cumulative Sum Charting

Another type of charting is the cumulative sum (cusum) control chart. The cusum chart has the advantage of identifying small persistent changes in the analytical scheme faster than the charts described above. The cusum chart may not be useful if one is just trying to evaluate individual results and is less interested in trends.

In the cusum chart system, each result is compared with a reference, usually the intended or expected value (17). The differences from the reference are then accumulated, respecting the sign, to give a cumulative sum of the differences from the standard. In other words, the difference between each result and the mean is calculated, and the second difference is added to the first, the third difference is added to the sum of the first and second, and so on. The cusum is plotted against time or analysis sequence to show the underlying level of the results. Its most important virtue is the damping out of the short-term variations, but highlighting of persistent or recurring differences. The cusum provides faster warning of process deviation from control and it offers a "look and see" approach to graphical estimation of trends (5). Some users of charts prefer the x-bar and range charts because they are easier to interpret.

Bicking and Gryna (18) in *Quality Control Handbook* provide instructions on how to prepare a "Cumulative Sum for the Average" chart, including the "mask" for interpretation of the angular change involved in the plot.

Recommendations

It is recommended that, under any quality assurance program, statistics be used to evaluate the quality of the data produced by the laboratory, the quality of the methods and chemical standards used by the laboratory, and the quality of overall performance. Procedures such as the sorting of data, the monitoring of analytical variance, and the use of such measures as standard deviation and relative standard deviation should be applied wherever they can supply insight into laboratory performance. Outlier values must be carefully considered as possible symptoms of analytical problems, and it is recommended that appropriate types of control charts be used to monitor the analytical processes.

References

(1) Albert, R. (1984) "Management Concepts in the Statistics of Quality Control of Laboratories," presented at an AOAC Quality Assurance Short Course, October 27–28, 1984, Washington, DC, Association of Official Analytical Chemists, Arlington, VA

(2) Daisley, P.A. (1986) "Basic Statistics — Applications in Quality Assurance," presented at an AOAC Quality Assurance Short Course, November 4–6, 1986, London, England, Association of Official Analytical Chemists, Arlington, VA

(3) The following may be found useful:

- *ASTM Manual on Presentation of Data and Control Charts Analysis* STP 15D, American Society for Testing and Materials, Philadelphia, PA
- Dewey, D.J., & Wilson, A.L., "Training Course on Analytical Quality Control for Water Analysis Laboratories," Water Research Center, Medmenham, England. This, in turn, is based on a more detailed publication: Cheeseman, R.V., & Wilson, A.L. (1978) *Manual on Analytical Quality Control for the Water Industry,* Technical Report TR66, Water Research Center, Medmenham, England
- Dixon, W.J. (1953) *Biometrics* **9**, 74
- Duncan, A.J. (1974) *Quality Control and Industrial Statistics,* 4th Ed., Richard D. Irwin, Inc., Homewood, IL
- Eckschlager, K. (1969) *Errors, Measurement, and Results in Chemical Analysis,* Van Nostrand Reinhold Co., New York, NY
- Grant, E.L., & Leavenworth, R.S. (1972) *Statistical Quality Control,* 4th Ed., McGraw-Hill, New York, NY
- Henry, J.A., & Knowler, L.A. (1980) *Two-Day Intensive Training Course in Elementary Statistical Quality Control,* 3rd Ed., American Society for Quality Control, Chicago, IL
- Natrella, M.G. (1963) *Experimental Statistics,* NBS Publication 91, National Institute of Standards and Technology, Gaithersburg, MD
- Snedecor, G.W., & Cochran, W.G. (1972) *Statistical Methods,* Iowa State University Press, Ames, IA
- "Standard Recommended Practice for Dealing with Outlying Observations" American Society for Testing and Materials, Philadelphia, PA
- Taylor, J.K. (1987) *Quality Assurance of Chemical Measurements,* Lewis Publishers, Inc., Chelsea, MI
- Wernimont, G.T. (1985) *Use of Statistics to Develop and Evaluate Analytical Methods,* Association of Official Analytical Chemists, Arlington, VA
- Youden, W.J. (1951) *Statistical Methods for Chemists,* John Wiley & Sons, New York, NY
- Youden, W.J., & Steiner, E.H. (1975) *Statistical Manual of the AOAC,* Association of Official Analytical Chemists, Arlington, VA

(4) Trout, C.R. (1986) "Statistical Sampling Approaches," presented at an AOAC-Agriculture Canada short course on "Field and Laboratory Sampling of Food and Agricultural Materials," March 25–26, 1986, Ottawa, Ontario, Canada, Association of Official Analytical Chemists, Arlington, VA

(5) *Chemistry Quality Assurance Handbook*, Vol. 1 (1982) U.S. Department of Agriculture, Food Safety and Inspection Service, Washington, DC, Sec. 1.5.13

(6) Kirchmer, C.J. (1988) "Basic Statistics: Applications in Quality Assurance," presented at an AOAC Quality Assurance Short Course, August 27–28, 1988, Palm Beach, FL, Association of Official Analytical Chemists, Arlington, VA

(7) Taylor, J.K. (1987) *Quality Assurance of Chemical Measurements*, Lewis Publishers, Inc., Chelsea, MI, p. 55

(8) Puri, S.C. (1981) *Statistical Aspects of Food Quality Assurance*, Information Services, Agriculture Canada, Ottawa, Ontario, Canada, p. 21

(9) Sprinthall, R.C. (1987) *Basic Statistical Analysis*, 2nd Ed., Prentice-Hall, Inc., Englewood Cliffs, NJ, p. 123

(10) Taylor, J.K. (1985) *Handbook for SRM Users*, NBS Publication 260-100, National Institute of Standards and Technology, Gaithersburg, MD, p. 78

(11) Benedetti-Pichler, A.A. (1956) *Essentials of Quantitative Analysis*, The Ronald Press Co., New York, NY, p. 11

(12) Wernimont, G.T. (1969) *Mat. Res. Stdu.* **9**, 8

(13) Statistical Technical Committee (1973) "Glossary and Tables for Statistical Quality Control," American Society for Quality Control, Milwaukee, WI

(14) McCully, K.A., & Lee, J.G. (1980) in *Optimizing Chemical Laboratory Performance Through the Application of Quality Assurance Principles*, F.M. Garfield et al. (Eds), Association of Official Analytical Chemists, Arlington, VA, p. 77

(15) Latimer, G.W. (1989) "Statistics and Control Charts," presented at an AOAC Quality Assurance Short Course, September 28–29, 1989, St. Louis, MO, Association of Official Analytical Chemists, Arlington, VA

(16) Westgard, J.O., et al. (1977) *Clin. Chem.* **23**, 1857

(17) Dux, J.P. (1986) *Handbook of Quality Assurance for the Analytical Chemistry Laboratory*, Van Nostrand Reinhold Co., New York, NY, p. 28

(18) Bicking, C.A., & Gryna, F.M. (1974) *Quality Control Handbook*, J.M. Juran et al. (Eds), McGraw-Hill Book Co., Newark, NJ, p. 23

Chapter 3
PERSONNEL CONSIDERATIONS

This chapter discusses ways in which the management of personnel and the application of recognized personnel principles can serve to improve and maintain quality. Among the most important responsibilities of management are those of planning, organizing, staffing, leading, and controlling (1), most of these directly or indirectly involving personnel management. For a laboratory director these duties involve setting short-term and long-term objectives; defining and establishing measures necessary to ensure the quality of results; organizing, managing, and motivating people; supervising technical performance; coordinating the efforts of individuals and groups; evaluating the success and failure of individuals, procedures, and programs and identifying the features that made the difference; and resolving everyday problems. The manager also, of course, has many other responsibilities, such as setting the hours of work, making work assignments, and protecting employee health and safety.

A high-quality personnel program will provide employees with the kinds of experiences that help them to maintain respect for themselves and their organization, and encourage them to set realistic goals and take responsibility for their fulfillment. Such a program will provide an environment in which employees will make judgments that are based on a rational perception of both their vital interests and those of the organization, and it will mobilize peer influence in support of a set of values that will be useful to the employees and to the laboratory.

It is well known that employee motivation is affected by such things as job design, promotion and pay policies, fair and timely evaluation of work, positive reaction to employees' suggestions, and participation in problem solving. It has been shown that good ideas that originate with the employees are more likely to succeed than are those imposed by management. This management style, practiced successfully by some Japanese and, as a result of their influence, some American companies, has been generally successful, demonstrating the achievement and maintenance of good quality and productivity when employees are allowed some genuine influence along with their traditional responsibilities.

The Role of The Laboratory Director

The performance of the laboratory director is key to a successful laboratory operation. Obviously the manager must have the appropriate education and experience to direct the scientific side of laboratory operations. But beyond this, the laboratory director must understand and be able to apply personnel management concepts and techniques in order to get the job done, and done well, by others. This includes knowing how to interact constructively with people and resolve the problems that management entails.

If an optimal level of laboratory efficiency is to be attained, the work that is to be done must be organized into jobs in a way that enables management to assign specific responsibilities to qualified people. The director must then establish position descriptions, recruit the appropriate personnel, place them in the proper positions, train them, evaluate their performance, and reward or discipline them as the situation demands. To ensure the laboratory's quality performance, he must direct the preparation of the quality assurance program, monitor its development and its implementation, and ensure its continued operation.

The Role of Supervisors

The first-line supervisor participates in a supporting position to the director in the management of the laboratory, including planning and policy formulation. The supervisor provides oversight and direction of the work flow to a group of scientists on a daily basis, offers guidance in the selection of equipment, keeps abreast of equipment preventive maintenance, ensures the proper validation of methods and results, reviews completed work and makes final conclusions on much of it, and may direct any participation in intralaboratory and interlaboratory proficiency programs.

The supervisor is in a critical position between the working scientists at the bench and upper management, and therefore must assume major responsibilities both toward the employees in the laboratory and toward the organization. On the one side, the supervisor plays an important role in the orientation and training of new people, and on the other the supervisor recommends to management appropriate hiring and training actions, including advanced training for more experienced staff members. The direct and continual appraisal of employee performance and the initiation of various personnel actions are among the other important functions of the supervisory position, making the first-line supervisor directly responsible for the administration and support of the quality assurance program.

The Responsibilities of Nonsupervisory Staff

Most accreditation programs and good laboratory practice regulations do not specify in any detail required qualifications and duties of professional personnel. The appropriate qualifications and experience of the scientific staff will, of course, depend on the requirements and responsibilities of each position. Obviously, each individual should be qualified to perform the assigned duties by virtue of successful educational or work experience, or a combination of the two. Rather than being specific with regard to professional qualifications, these programs and regulations strongly recommend the maintenance of job descriptions, records of training and work experience, and records of other related personnel actions in a suitable personnel file.

As a minimum, each analyst must understand the importance of the quality assurance program and the importance of the role of every employee in making it succeed. The employee should recognize potential sources of error in assigned tasks, report situations in which the quality of work falls below expected levels, abide by safety and housekeeping rules, and use good judgment as well as some level of thoughtfulness or ingenuity in the performance of day-to-day assignments. In particular each analyst must understand the principles of methods used, follow methods as written or carefully document any deviations, keep accurate records, and have a basic knowledge of statistics and its application in laboratory quality control.

Most laboratories employ helpers, aides, and technicians who work under the supervision of operating professional personnel. With adequate indoctrination and training, these employees should be capable of performing a number of essential laboratory activities such as washing laboratory glassware, servicing equipment, preparing samples for analysis, preparing various types of solutions, and even conducting noncomplex analyses. As a minimum, each of these nonprofessionals must understand the assignment, the importance of performing duties at the necessary level to produce quality results, and the necessity to report unexpected observations that may indicate situations that require the attention of a professional analyst. Any analyst who has been victimized by poorly cleaned glassware or the unavailability of a routinely stocked reagent knows very well the importance of the role of the nonprofessional. It is imperative that the nonprofessionals know of that importance themselves.

Position Qualifications and Position Descriptions

Most organizations have rather well-developed qualification requirements for various scientific positions. This element of personnel management is important for recruitment, selection of personnel, and proper compensation. A position qualification can be defined as any quality, knowledge, ability, experience, or required attribute that contributes significantly to a person's ability to perform satisfactorily in a particular position. The criteria generally used are education, experience, special skills, responsibility to interact with others, and the level of supervision the person requires. Once these attributes are established and coupled with the job duties described in the position description, the position can be evaluated with respect to level of pay. The mechanism is actually quite complex and, unless it is handled properly, can lead to salary inequities and serious employee dissatisfaction.

The format for position descriptions varies, but it must include a summary description of the position, a general list of duties, and the level of supervision and guidance received. Some organizations include standards of performance in the position descriptions, but if they are not listed, they should follow as a separate document soon after employment commences. The standards state how well each duty is to be performed, and the first-line supervisor must explain clearly and carefully exactly how that performance will be measured and discussed. Each individual standard should be as specific as possible and stress the results to be achieved. These performance standards have significant value because they provide the employee and the supervisor with a tool to evaluate the quantity, quality, consistency, and timeliness of work performance. The benefits derived are worth the additional efforts required to develop and implement a performance standard system, preferably as a part of the position descriptions.

The Preemployment Interview

Laboratory quality may be said to begin with the preemployment interview, where the prospective employee must be made to understand the rigid requirements with respect to quality that the job demands, and the prospective employer must use quality interviewing techniques as well as quality judgment to make appropriate selections from among candidates. Every information source available to the recruiter must be used, and this includes at least the preemployment interview, the information in the application, discussions with previous employers, and school transcripts. The recruiter should ask questions that require the candidate to give full answers, rather than a yes or no. Technical questions might relate to problems encountered or solved in school or with previous employers. The person's aspirations can be questioned: for example, "Why do you want to work in this laboratory?" or "Where do you expect to be five years from now?" Facility with verbal expression can be assessed from both the interview and the quality of the completed job application form itself, particularly if it was filled out on site. It is essential, of course, that questions relate to the ability to perform the job, and not to extraneous personal matters.

Orientation and Training

All new employees normally receive some form of orientation once they report for duty. Programs vary, ranging from brief informal introductions to lengthy planned discussions. The extent and type of information sought varies with the level of the position. The information imparted usually involves matters of immediate concern, such as work hours, work conditions, pay periods, personnel policies, benefits, safety requirements, organizational structure, introduction to co-workers, and other items that influence work and welfare. For laboratory personnel the program must extend to more technical matters to provide a basic understanding of the role of the laboratory and some of the basic elements of the job itself.

This is an important opportunity to explain the laboratory quality assurance program and its significance. It is important that this program be presented to the new employee immediately, and that every effort be made to emphasize the fact that the program is of major importance to top management in the organization and that it is expected that it will be treated with importance by employees at every level in the organization.

There may not be a sharp distinction between orientation and the on-the-job training program; the two actually go hand in hand. The kind of thing that should be avoided is handing the new employee a large package of reading material, and considering that to be "training." Orientation and training must be a one-on-one situation, with close monitoring by a supervisor or experienced professional so that the new employee will learn and learn correctly, not develop a feeling that no one is interested and that doing a good job is probably not that important anyway.

On-the-job training—working directly with professionals who know the job well and can communicate its features to the new employee—is essential. It helps provide knowledge and the technical skills that are necessary to perform the tasks in the job description, and it stresses the importance of quality performance.

Horwitz et al. (2) summarize the training of new analysts and the concept of quality control in a statement that is quite appropriate for an on-the-job training program:

> Many laboratories find that they must institute training programs and on-the-job training for their new analysts, particularly recent college graduates. Training usually ends for each type of determination when reproducible results are routinely obtained. The analyst is then considered fully trained and is placed in the normal stream of the laboratory workload. By constructing and maintaining control charts of replicates, some laboratories have utilized the elementary concepts of quality control to assess the progress of training of new analysts. After the replicates are in control, the analyst is required to maintain the control chart by performing duplicate assays on a predetermined fraction of samples, say five percent, if single determinations are usually run. Other procedures, mentioned previously, can also be used for this purpose — reanalyzing stable previously analyzed samples, testing fortified or standard samples, and assaying samples analyzed in another laboratory or by other analysts. The result will be a practical demonstration of the ability of analysts to maintain performance within personally developed control limits. Lapses indicate a need to ascertain the reason rather than a cause for reprimand.
>
> As each new method and its application to various samples are brought under control, each analyst supplies his or her own evidence of mastery, ability to apply training in a routine fashion, and readiness for the next step in acquiring journeyman proficiency. In this manner, subjective judgments are avoided, and concrete evidence of proficiency is obtained; and the evidence is acquired with the wholehearted cooperation of the trainee, rather than with an attitude of resentfulness and wanting to get on with the job.

Training is also used to redress discovered deficiencies in performance, to learn more about a new or intricate technique, to understand a new program, to see whether a new analytical method can solve a problem, to learn how to use an instrument, and to prepare a person for future responsibilities through career development.

Career development training is generally aimed at increasing the long-term effectiveness of employees in the light of the organization's projected needs. The analysis of long-term training needs is complex. It requires attention to the employee's abilities, interests, and aspirations and the future plans, programs, and goals of the organization. These goals must be integrated to obtain major benefits from the training. It is desirable to develop a separate career development agenda for each employee, but this is frequently overlooked and training is approached in an *ad hoc* manner.

Training Methods

On-the-job training, described above, has as one of its purposes the development of skills and the introduction to local procedures. Others are the formation of attitudes toward and identification with the position and the organization and to ready the person to take on assigned responsibilities.

Another training procedure is **individual instruction,** where the employee is required to study written materials such as the quality assurance manual or a methods manual or other documents to prepare for a particular assignment or area of operation. Among the materials in this category are **programmed instruction** courses, a form of self-teaching that has been found to be quite effective. In a large organization, training materials can be generated in-house.

When training groups can be brought together, lectures or **classroom training** are feasible. The advantages of this approach are that it can be used to reach a larger number of people at one time, and there is some assurance of the consistency of instruction among the trainees. It is an effective way to introduce the quality assurance program as well as other material that has general applicability to the group, such as the safety program. The instructors must not only be competent in the subject matter, but have proven ability to communicate in a classroom environment. An amateurish approach could do more harm than good because it may appear that the organization does not take the subject matter seriously, but simply wants to discharge an obligation.

Over the past 5–10 years, a number of training products have become available on the market that use **audiovisual materials** for the presentation of information to individuals or groups. These materials include audio tapes, slides, and video tapes. There are several professional organizations, and some private ones, that rent or sell these materials.

Scientific associations and organizations conduct **short courses** that can be of value, although the brevity of the courses and the costs for registration and travel must be taken into consideration. The American Chemical Society and the Association of Official Analytical Chemists are among the scientific organizations that present such courses, usually in conjunction with their meetings.

Attendance at **scientific meetings** offers opportunities to learn of new advances in a technical field, new methods, and new techniques. Also important is the contact that a scientist makes with peers working in other organizations.

University and college training is certainly a highly recommended source for further education. Review courses, seminars, and continuing education programs are offered at almost all educational institutions in a variety of disciplines for full-time or evening attendance. This would be a reasonable place to look for sound training in statistics.

Specialized training by **instrument vendors** is widely available but must be approached cautiously since it is more often "training" in what to buy rather than in principles of operation.

Finally, much can be learned from **visits to other laboratories,** where sufficient time is available to discuss new procedures and techniques with peers. Such visits also establish contacts that can lead to future discussions or training exchanges.

Performance Appraisals

The continuing appraisal of employee performance is an essential element to a functional quality assurance system, yet formal appraisals are often neglected or performed poorly. Most supervisors and subordinates approach the appraisal session with dread and trepidation, but if they both understand the purpose of appraisal, and the supervisor is objective and gives serious thought to the process, most of the fears on both sides will be allayed.

Rhetoric aside, the formal appraisal is simply an attempt to think and then communicate clearly about each person's performance and the prospects for advancement against a background of the total work situation. It is also an opportunity to improve or renew communication between the supervisor and the employee. Improving job performance is the primary objective of the formal appraisal system, but it has a much broader effect, such as motivating employees by providing frank and direct information on how they are doing or could improve, and by assisting supervisors in making decisions concerning training and development needs and staffing. If the formal appraisal is constructive, the employee will recognize that something worthwhile has been accomplished. The effect on the employee's career will be obvious and will encourage a healthier, more positive attitude about the job and chances for advancement. The supervisor, on the other hand, may develop a better understanding of the employee and an appreciation of the employee's strengths as well as weaknesses. The supervisor may even see new ways to handle long-standing or vaguely recognized problems.

The supervisor who feels that day-to-day contacts with employees can serve to evaluate their performance is wrong. It is true that appraisal is a process that should be happening continually — it is not a once or twice a year event. But there is no doubt about the fact that a formal "taking stock" is necessary to bring home to both the employee and the supervisor exactly where each stands with respect to job performance, to make definite plans, if necessary, to improve job performance, and to have serious, unhurried discussions on career plans and laboratory needs.

Appraisal interviews with employees are held once or twice a year and more frequently with new people in an intensive training program. The following subjects are typical of ones considered during appraisal sessions (3): performance objectives, quality and productivity requirements, work safety, top-priority job elements, classification of responsibilities, personal long-range goals, barriers to good performance and their causes and cures, and other topics of importance to either party at the time. An immediate goal is to set in motion an agreed-upon plan that will help the employee to improve effectiveness on the job.

Many companies use a rating form to evaluate personnel performance. Federal agencies, for example, use a system identified as "Employee Performance Management System," which consists of performance elements, with quality standards for each. An element describes a particular type of performance (for example, productivity) in somewhat quantitative terms. The quality standards describe what kind of performance partially meets, fully meets, or exceeds the requirement for the element. The performance also can be rated below this scale (fails to meet) or above this scale (substantially exceeds). The kinds of elements that are rated include quantity of work performed, appropriateness of work relative to grade, timeliness (meeting deadlines), quality, research output in some cases, conformance with safety and other regulations, and efforts to keep current in the employee's field.

Self-Evaluation

Taylor (4) suggests the use of a "Quality Assurance Training Profile," a checklist that can be used by staff members as a means of self-appraisal or used by supervisors in making decisions about training needs. Ten elements are considered and a numerical scoring system is provided to rate the appraisal from "deficient" to "outstanding." Nine of the elements represent the individual's knowledge and experience in general analytical chemistry, specific field of activity, statistics, control charting, standardization activities, quality assurance, sampling, calibration, and data evaluation. The 10th element is skill in reporting and report writing.

Taylor also suggests the use of a second, similar, self-appraisal checklist titled "Personal Quality Assurance Profile." Fifteen items are listed on this form, ranging from "Knowledge of Field" to "Housekeeping Practices." This form also contains a scoring system and some criteria for judging acceptability. The other items on this form are General Understanding of Methods Used, Mastery of Specific Technology, Use of Written Standard Operating Procedures, Pre-check of Methods Prior to Use, Adherence to Good Laboratory Practices and Good Measurement Practices, Laboratory Notebooks, Knowledge of Statistics, Control Chart Usage, Participation in Technical Activities, Training Courses, Technical Books Read and Informal Training Received, Experimental Planning, and Use of Randomization. This is an interesting and excellent approach to self-evaluation and performance improvement.

The Personnel File

An official personnel file is necessary for each employee. All personnel matters pertaining to the employee, such as position descriptions, training evaluations, participation in proficiency test sample programs, promotions, awards, and interviews, are kept in the file. It should be updated as necessary and transferred with the employee from location to location in the organization. The file serves several purposes: for example, it can be used to determine what additional training the employee should receive, how the employee fares in comparison with others in proficiency tests, whether or not the employee's efficiency and work habits would support promotion, and so forth. When a laboratory seeks accreditation by an accrediting body, the auditor who visits the laboratory will invariably ask to see individual personnel files in order to establish the ability of the laboratory to conduct tests in the area in which it seeks accreditation. Lack of information on individuals is likely to result in rejection of the request for accreditation.

Since some of the information in the file may be sensitive and personal, and subject to privacy protection regulations, access to the file should be limited, and a record kept of people who draw the file for any purpose.

Recommendations

It is recommended that the following steps be taken in the development of the personnel management program:

1. Each laboratory person be given a general understanding of personnel and employee development policy and be apprised of any specific responsibilities.
2. Jobs be analyzed, and qualification requirements, salary, position descriptions, and performance standards be developed for each position.
3. Recruiters be trained in interview techniques and applicant evaluation.
4. Orientation and training programs for new employees be developed and their use documented.
5. Short- and long-range training plans for the development and advancement of laboratory personnel be prepared.
6. A program for formal employee appraisal, including evaluation criteria, and a suitable schedule and format for the appraisal interview be developed.
7. Personnel files that contain information on education, experience, salary, training, promotions, appraisal, and other personnel actions be maintained.
8. Employees be encouraged to evaluate their own performance and training needs.

References

(1) Dessler, G. (1982) *Management Fundamentals, Modern Principles, and Practices,* Preston Publishing Co., Inc., Reston, VA, p. 3
(2) Horwitz, W., et al. (1977) in *Quality Assurance Practices for Health Laboratories,* S.L. Inhorn (Ed.), American Public Health Association, Washington, DC, p. 640
(3) Mayfield, H. (March–April 1960) "In Defense of Performance Appraisal," *Harvard Business Review*
(4) Taylor, J.K. (1987) *Quality Assurance of Chemical Measurements,* Lewis Publishers, Chelsea, MI, pp. 177, 185

Chapter 4

MANAGEMENT OF EQUIPMENT AND SUPPLIES

Equipment Selection and Purchase

There can be no satisfactory quality assurance without adequate and planned attention to equipment management. Advances in the design of electronic analytical instruments have served to improve the quality of analytical measurements and increase laboratory productivity. The introduction of computers has brought further improvements, their use permitting unprecedented interpretation of data.

Because these instruments are expensive and their characteristics vary from manufacturer to manufacturer, care is necessary in instrument selection and installation. Many laboratory requirements need to be considered, such as workload volume, experience of the staff, accuracy and resolution required, and the ruggedness needed based on expected ability to purchase replacements. In addition to cost and performance characteristics, then, the purchaser must consider the length and conditions of warranties, expected downtime, the prompt availability of repairs and the costs involved, supplier-offered training, and whether or not the supplier will install and test the item against specifications. Consideration also should be given to standardizing on one vendor's instruments for the convenience of interchangeable parts and single-source servicing. Important information on instrument operation can be obtained while attending equipment exhibits and demonstrations at scientific meetings. Some suppliers will furnish equipment on a temporary basis for trial use and evaluation, and most will provide demonstrations either at the customer's laboratory or at their own facility. Every advantage must be taken of these possibilities to ensure that the eventual purchase is based upon reliable information and a good comparative evaluation of available models.

Other items can be considered during negotiations for the purchase of analytical instruments: for example, a fixed date for delivery and complete installation; adequate training, on site or at the manufacturer's facility, not only for day-to-day users but for one or two fully experienced monitors as well as (in some cases) an electronics technician; extended warranties or service contracts; 2 copies of all operating and service manuals that are available; sufficient supplies and spare parts to ensure immediate startup and continued operation through a break-in evaluation period; and auxiliary items such as spectra libraries, specialty columns, or interfacing with other available equipment.

Equipment Installation and Servicing

Before new equipment is purchased, serious thought must be given to its location and installation. For most instruments, it is not advisable to move them more often than absolutely necessary, so not only a satisfactory, but a relatively permanent, location should be sought. A full understanding of the requirements of the new equipment has to be obtained well in advance from the vendor: for example, required bench or floor space and in some cases bench height; utility needs such as water, electricity, compressed gases, or lighting; security considerations; and a compatible environment. Whether the equipment is installed by laboratory personnel, by the vendor, or under an outside service contract, it will require prompt checking against specifications and performance requirements. If the item does not perform as required, it cannot be placed into service, and therefore should not be given final approval. In such a case, consideration may be given to withholding final payment until a correction is made by the supplier.

Wiring diagrams and significant technical details often will not accompany instruments if inclusion of this material was not required as a condition of purchase, as suggested above. When obtained, the diagrams, manuals, and charts should be studied by the monitor or electronics technician and kept in a safe place for reference purposes. If 2 copies are obtained, one should be secured in the event the working copy is lost or damaged. Often, equipment will outlive the availability of its manuals.

The increased use of solid-state electronics in instruments places them beyond the understanding of most analysts to troubleshoot and repair. Therefore, when instruments break down or fail to function properly it is generally necessary to call in instrument repair experts, which can mean delays in analyses and, in many cases, considerable expense. There is no easy, or universal, solution to this problem. Some suppliers offer routine maintenance training as part of the purchase contract, some of which may go to a laboratory electronics technician. The use of outside service contracts is also a possibility. As suggested above, "deals" can often be made at the time of equipment purchase, when suppliers are most likely to be amenable to throwing in "extras" that could cover servicing for a period of time. In any event, an important cost-saving step is to identify as soon as possible parts that are likely to need periodic replacement, and then maintain a stock of these parts. Experience will allow the monitor or electronics technician to add to this list those items that tend to malfunction and that also are known to have substantial shelf life.

Once a piece of equipment is approved for use, it should be entered into the laboratory's inventory and to the schedule for preventive maintenance.

Preventive Maintenance for Equipment

Although state-of-the-art instruments are important additions to the laboratory analytical system, periodic servicing and calibration are an absolute necessity. This type of preventive maintenance reduces malfunctions, permits adjustments on a timely basis, and most important, ensures fewer and shorter equipment breakdowns and increased measurement system reliability.

Preventive maintenance includes specification checks, calibration, cleaning, lubricating, reconditioning, adjusting, and testing. It is essential that calibration standards be maintained and performance over time be compared and analyzed. A need for troubleshooting is indicated when it is evident that a shift in instrument response has occurred.

The preventive maintenance program needs to be approached in an organized and integrated manner, bearing in mind its importance in the quality assurance system. A number of steps are required in establishing the program. A nine-point scheme is described below that has been used successfully by a number of laboratories (1). Modification of the suggested scheme is, of course, possible, and other approaches can be adapted to fit particular situations.

1. *Inventory.* A permanent inventory record is established for each piece of equipment that costs more than an established minimum price. The record includes name of the item, model number, serial number, vendor's name and address, date received, date checked against specifications, date placed in service, cost, and laboratory location. This system lends itself to computerization and thus reduces much checking and retyping when inventory time rolls around.

2. *Definition of Service Tasks.* A quality assurance performance checklist is maintained for each piece of equipment in a bound notebook kept near the equipment or in a common location with other similar notebooks. Information recorded includes name of the item, frequency of check or necessary services, standard reference material or other calibration materials to be used, and general procedures to be followed. Many laboratories also include a service record to cover downtime, performance problems, repairs and cost of repairs, date of breakdown, date of return to service, who made the repairs, and parts required. This information serves a number of purposes and is especially helpful in deciding whether or not to retain a given supplier, when the equipment is ready for replacement, or when there is a need for additional units to be purchased. A "Repair Control Form" can be used instead of recording repair in the preventive maintenance notebook. Some laboratories include this repair information in their computerized equipment inventory system.

3. *Interval Establishment.* The frequency with which the service and calibration tasks are to be performed is established and recorded. The frequency will depend on the type of equipment; the supplier's recommendation; maintenance and servicing history; extent and severity of use; age; tendency of the

item to wear or drift; environmental conditions such as ambient temperature, humidity, and vibration; and the quality of measurements sought. When internal standards or system suitability tests are used, for example in chromatographic procedures, the usual resolution and reproducibility measurements are performed to determine instrument stability at the time of analysis making it unnecessary to establish other periodic performance checks (2). Regardless of how the checks are made, the information, the name of the person who made the checks, and the date are recorded in the notebook.

4. *Personal Assignment Monitors.* These individuals and their alternates are assigned preventive maintenance duties for each piece of equipment. This is to ensure that the tasks are performed as scheduled. An alternative procedure used in some laboratories is for a supervisor, on a scheduled basis, to assign someone who is available and capable. Obviously such assignments should be commensurate with aptitude, training, and experience. For some instruments, it may be desirable to provide regular contract servicing in addition to in-house activities.

5. *Special Instructions.* For some instruments, special monitoring devices, techniques, materials, or a special piece of equipment may be necessary to check the performance. These special items are to be listed in the notebook and their use made a part of the maintenance for that particular instrument.

6. *Training.* A program is established to train personnel in the performance of the more difficult maintenance and repair tasks. As mentioned earlier, suppliers may be willing to furnish the training.

7. *Operating the System.* To make certain that the preventive maintenance jobs are performed as scheduled, a reminder scheme is established to notify monitors, in advance, of their obligations. This system can be as simple as a wall chart, or it can be a computer program, or special written assignments prepared by the laboratory's clerical staff under some form of "tickler" follow-up system.

8. *Records and Documentation.* A record is made in the appropriate notebook documenting completion of the maintenance tasks. Any equipment deficiency is recorded, and if serious, reported to a supervisor so that corrective action can be taken. When correction is made, it is recorded with details of what was done.

9. *Surveillance.* A scheduled periodic review of notebooks is made by a supervisor or quality assurance coordinator to ensure that the various preventive maintenance tasks have been performed, and this review also is recorded.

It is not feasible in this text to detail a preventive maintenance schedule that will cover all pieces of equipment in general use, or one that can be adopted universally by all laboratories. Frequent introductions of new instruments and instrument modifications make this task virtually impossible. Appendix C, "Instrument Performance Checks," provides one excellent set of criteria taken from the *Laboratory Quality Assurance Manual* of the Field Operations Directorate, Health Protection Branch, Health and Welfare Canada, Ottawa.

An American Public Health Association book, *Quality Assurances Practices for Health Laboratories* (*3*), also contains suggestions in some detail in Chapter 1 under "Maintenance of Common Laboratory Equipment," and in a table in Chapter 11 identified as "Maintenance Criteria and Performance Schedules for Instruments Used in the Food Laboratory." For laboratories that do not have established equipment performance criteria, review of these references can be helpful for initiating a suitable program.

An important part of equipment management is a plan for capital equipment acquisition and replacement. To a large degree such a plan will depend upon information on the performance suitability, downtime, and repair history of current equipment. The purchase of equipment that will not be fully used, or the continued use of obsolete equipment, or the expenditure of funds on the continual repair of items that would better be replaced, is wasteful of funds as well as personnel time, but data on performance is needed to demonstrate this. Also of considerable importance is the fact that the failure to procure new state-of-the-art equipment can severely limit the laboratory's technical competence.

Supply Management

Quality work depends in part on having appropriate quality reagents, volumetric glassware, and other supplies on hand when they are needed. The use of suitable procurement specifications and good inventory procedures will help to ensure this. Reagent specifications must take into consideration identity and purity. A good inventory procedure will have to deal with vendor source, catalog numbers, container sizes, grades (as defined by each vendor), safety, stability, storage and handling requirements, and reorder points. Some computerized inventory systems will automatically provide lists of items that are needed to establish the stock at a given level above the reorder point, and some will even prepare purchase requisitions for those items. Whether a sophisticated computerized inventory is used, or a hand-written tally, the laboratory must be able to anticipate shortages in sufficient time to avoid having to experience them.

Laboratory personnel must understand their responsibilities in the use of reagents, solvents, standards, and glassware of appropriate quality for the analyses they conduct. For example, a reagent fit for use in a drug analysis where the drug entity is the principle ingredient, may not be proper in a pesticide determination where the analyte is present at the parts-per-million level.

Chemicals

Chemical reagents, solvents, and gases are generally available in various grades and levels of purity. Reagents are available in about six grades with purity specifications being the principal difference among them (*4*):

- Primary standards — Each lot is analyzed and the purity specified.

- Analyzed reagents — These fall into 2 classes: (a) each lot is analyzed and the percentage of impurities reported, and (b) conformity with specified tolerances is claimed or the maximum level of impurities is listed. Analytical reagent grade, reagent grade, and ACS analytical grade are examples. Their terms appear to be synonymous, and reagents so labeled conform with the specifications of the Committee on Analytical Reagents of the American Chemical Society.
- USP Reference Standards — These are chemical reference standards for use as control or comparison standards in certain official assays and tests for compendial articles in the *United States Pharmacopeia*. They are purified specimens of commercial drug substances for which manufacturing and purifications processes are well established. Each lot is tested for purity prior to issuance as a standard by USP or a cooperating laboratory, but regardless of the level of purity achieved, these materials are standards by definition when used as directed in USP monographs.
- Pure, C.P., Chemically Pure, Highest Purity — These are qualitative statements used by some chemical manufacturers that are without universally specific meaning.
- Purified, Practical Grade — These are usually bulk chemical starting substances used for synthesis or to be subjected to further purification.
- Technical Grade, Commercial Grade — These are chemicals of widely varying purity.

The first three classes noted above can be used for most analytical work without further testing, bearing in mind that special types of analyses may require further special treatment of the reagents before use. It is good practice to check new lots of reagents for performance against previously satisfactory lots.

Special attention must be given to reagents, solvents, chromatographic materials, and, when testing for materials in the parts-per-million range, even to filter paper. Some methods may include specific preparation requirements for reagents, such as distillation, regardless of the purity claims on the reagent label. If interferences are observed when these materials are used, special purification steps may be required. Improper or long-term storage may lead to problems with some solvents even if they were initially found to be satisfactory. Use of a method blank with each group of samples being analyzed may be useful to detect such interferences, or to provide evidence that they do not exist. If the method blank indicates the presence of potential interfering substances, then a reagent blank or item-by-item reagent testing may be needed to establish the cause of the analytical problem.

Normally, reagents will be purchased in bottles of such size that they will be completely used within a reasonable period of time, reducing the possibility of quality deterioration. A wise step is to place a receipt date on each bottle label, and an "opened" date, initialed by the analyst, to establish the age of the reagent. In certain cases it may be important to add an expiration date, for example, for ethers. Obviously, all reagent containers must be fully labeled and tightly closed, and any "leftovers" discarded rather than being returned to original bottles.

Reference Standards

Much of the material in this section refers to publications from the National Institute of Standards and Technology (5) and the United States Pharmacopeia (6).

A *Reference Material (RM)* is a material or substance, one or more properties of which are sufficiently well established for it to be used for the calibration of an apparatus, the assessment of a measurement method, or for assigning values to materials (7).

A *Certified Reference Material (CRM)* is a material, one or more of whose property values are certified by a technically valid procedure, accompanied by or traceable to a certificate or other documentation issued by a certifying body.

A *Standard Reference Material (SRM)* is a material issued by the U.S National Bureau of Standards, which was recently renamed National Institute for Standards and Technology (NIST). SRMs are certified for specific chemical or physical properties and are issued with certificates that report the results of the characterization and indicate the use of the material. These are well-characterized materials produced in quantity to improve measurement science.

A *Reference Material (RM)* listed in "NBS Standard Reference Materials Catalog" is sold by, but not certified by NIST. They meet the International Organization for Standardization (ISO) definition for RMs and many meet the definition for CRMs. NIST offers for sale over 900 different materials through its Office of Standard Reference Materials. All materials bear distinguishing names and numbers by which they are permanently identified.

The New Brunswick Laboratory of the U.S Department of Energy issues special nuclear reference materials as *NBL Certified Reference Materials.* These materials were issued by the National Bureau of Standards before October 1, 1987, but are now available from other sources. The ISO, through its Council Committee on Reference Materials (REMCO), has prepared an International Directory of Certified Reference Materials. Inquiries may be directed to the address shown in the reference at the end of the chapter.

The International Union of Pure and Applied Chemistry (IUPAC), through its Commission on Physicochemical Measurements and Standards, issues a catalog of CRMs that are useful for physicochemical properties.

RMs and CRMs are quite expensive and their use is generally limited to the calibration of instruments and methods in order to establish long-term reliability and integrity for accuracy and resolution. These standards are also used to assay or calibrate so-called working or secondary standards that are used as reference materials in day-to-day analyses. This secondary standard material then becomes the laboratorys "house" reference material or working standard.

The United States Pharmacopeia (USP) has established a collection of some 800 reference materials that cover more than 90% of the drugs that are likely candidates for testing. *USP Reference Standards* for compendial articles are purified specimens of commercial drug substances for which manufacturing and purification processes are established. A list of current lots of USP Reference standards is published periodically in the USPs *Pharmacopeial Forum,* available from Pharmacopeial Convention, Inc, 12601 Twinbrook Parkway, Rockville, MD 20852.

The Association of Official Analytical Chemists, in its *Official Methods of Analysis,* lists a number of U.S. and foreign national sources of certified reference materials for animal tissues, plant tissues, foods, alcoholic beverages, animal feedstuffs, biochemicals, chemicals, drugs, industrial hygiene materials, fertilizers and related materials, pesticides, water contaminants, sediments, gases, particulates, and fuels (*8*).

Four guides related to reference materials have been issued by the ISO. Guide 30 (1981) covers terms related to materials, measurements and testing, methods, certification, and issuance. ISO Guide 31 (1981) addresses contents of certificates, what should be stated about every reference material, and the certified values and confidence limits. ISO Guide 33 (1981, updated 1989) is a user's guide for proper selection and use of reference materials, and includes some information on misuse, scope, definitions, measurement processes, and applicability. Finally, ISO Guide 35 (1989) concerns the certification of reference materials, principles, homogeneity, definitive methods, and the uncertainty of measurement.

Regardless of the initial purity of reference materials, care is necessary to see that they are packaged, stored, and handled to prevent deterioration. Precautions are required to minimize exposure to moisture, air, heat, and light, the primary causes of deterioration. To ensure the integrity of the standards they must be secured, and records maintained of their receipt and use.

A reasonable approach to the control of standards is to assign monitors to such duties as ordering, refilling empty containers of working standards, properly identifying containers, maintaining standards in their proper locations, disposing of old or outdated standards, maintaining standards up to date, and checking calculations of standards assays when conducted. Standards records should be kept in sign-in/sign-out logbooks located in the standards storage areas. Each analyst using a standard would be required to sign an entry in the logbook that indicates the name of the standard, and the date and time it is taken and returned. For controlled materials, the weighings that indicate the amount of material taken and a notation concerning the purpose (such as particular sample numbers) may be required for security reasons. Each analyst must be made to appreciate the importance of standards and their proper care.

House reference materials or working standards are to be assayed by the best method available and the results entered in a notebook kept for the purpose, along with the analyst's name, date of analysis, source and lot number of the material, all raw data, charts, calculations, and so forth. The house material is handled in the same manner as RMs. When the house material is used in conjunction with an assay of a sample, reference to it is reported with the sample results.

Standard Solutions

All chemistry laboratories maintain stocks of standard solutions. The number, kind, and amounts depend on the needs of the particular laboratory, including the frequency of use. Improper standardization has been found to be one of the leading causes of error in the analytical scheme. Errors are usually caused by mistakes in calculations, standardization for one purpose and use for another (for example, the interchange of methyl orange and phenolphthalein indicators in acid-base titrations), standardization at one temperature and use at another (especially for nonaqueous solutions), concentration change due to solvent loss, and deterioration on storage (carbon dioxide absorption by sodium hydroxide standard solutions, or the growth of mold in thiosulfate solutions, for example). The best practice is to standardize solutions at the time the assay is performed.

When stock solutions are to be maintained for general use, several precautions and special procedures are advisable: (a) Have the preparation of solutions and their standardization assigned to specific monitors; (b) have the monitors maintain records in suitable notebooks that indicate the identity, concentration, method of preparation, standardization procedure, calculations, date, and monitor's name; (c) see that a check standardization is conducted by a second analyst; and (d) have a label placed on the container showing essential information. In some instances, especially when experience has shown that the solution does not retain its standard value over an extended period, it is important that an expiration date be added to the container label. A paper by Lam and Isenhour, titled "Minimum Relative Error in the Preparation of Standard Solutions by Judicious Choice of Volumetric Glassware," contains valuable information on standard solutions (9).

Purified Water

Purified water is one of the most critical, but one of the most often neglected, reagents used in the laboratory, in spite of the fact that all analyses are affected by the quality of the water used. Distillation of water will not always ensure its quality. The design of the still, the materials of construction, the character of the raw water, and the rate of distillation can all influence the quality of the distillate. Stills need periodic cleaning to remove scale. When the feedwater is of poor quality because of hardness and dissolved organic compounds, it may be necessary to combine a water softening and carbon filtration system before distillation. For metals analyses, it may be necessary to use water redistilled from an all-Pyrex glass distillation system. Special clean water systems, such as Millipore/Milli-Q, may be necessary for trace analysis. The storage container, too, can significantly affect the quality of water, especially if the water is stored for extended periods.

A high-grade ion exchange system can produce water of acceptable quality for many purposes, but this method does not remove certain impurities. Specific conductance or specific resistance is used as a measure of inorganic contaminants in water. Purified water, or ACS reagent grade water, can be defined as water that has been distilled or deionized so that it has a specific resistance of more than 500,000 ohms or a conductivity of 2.0 micromhos. ACS specifications for distilled water for general analytical use are (10):

Residue after evaporation	Not more than 1 mg/L
Chloride and ammonia	Not more than 0.1 mg/L
Heavy metals (as lead)	Not more than 0.01 mg/L
Permanganate test for organics	Color persists 1 hr at room temperature

The permanganate test involves the addition of 0.03 mL of 0.1 N KMnO$_4$ to a mixture of 500 mL sample and 1 mL H$_2$SO$_4$.

The ASTM has established specification for four grades of purified water ranging from Type 1, reagent water to be used where maximum accuracy and precision are necessary, to Type IV, reagent water of moderate purity that can be used in procedures that require large amounts of water (11). The USP also defines several water grades, including high purity, purified water, and water for injection.

The American Public Health Association (APHA) suggests that potable tapwater be tested at least semi-annually by the coliform membrane filter procedure described in its standard (12). Either distilled or demineralized water can be used for the preparation of media and reagents for microbiological use, but it is recommended that quality control tests be conducted regularly to establish the potential toxicity or prior contamination of this type of laboratory water. A distilled water suitability test is described in APHA's *Standard Methods*. Because distilled water may contain chlorine, it is suggested that sodium sulfite or sodium thiosulfate be used to neutralize the chlorine before distillation. Suitable records are to be maintained to demonstrate that the quality control tests were conducted.

In summary, significant attention must be given to the purity of water used in the analytical laboratory, from the quality of the starting raw water, to the system used to purify the water, the manner used to deliver the water, the containers used to store it, the tests used to establish the quality of the purified water, and certainly to the purposes for which the water is to be used.

Culture Media

Many media formulations are available commercially in either dehydrated or ready-to-use form. They have been found to be suitable for use in methods presented in the *Food and Drug Administration Bacteriological Analytical Manual,* referred to as the *BAM*, published and distributed by the AOAC. It is important that the instructions provided by the media manufacturers for preparation and use be followed. According to the *BAM*, the quality of commercial dehydrated media is generally controlled better than corresponding media formulated in the average laboratory kitchen. The *BAM* also suggests that it be a general practice to inoculate a set of control organisms on each lot of medium, especially for laboratories engaged in critical research or the development of new methods that depend on specific biochemical characterization of organisms.

The *BAM* lists some 60 pages of instructions for culture media preparation for laboratories that prepare their own. Appendix B contains instructions for the preparation of reagents and diluents, and Appendix C has instructions for stains and staining procedures.

Volumetric Glassware

Volumetric measurements are an essential element in an analytical laboratory since so many determinations require specific dilutions and controlled delivery of various amounts of accurately prepared solutions. Analysts too often take at face value the volume markings on flasks, pipets, and burets, without actually establishing the true accuracy. Similar assumptions are made about volumetric syringes used in chromatography.

A number of extraneous conditions can also influence the precision of a given measurement. They include temperature, method of reading, cleanliness of the internal glassware surface, method of delivery, depth of color of the solution, type of meniscus, calibration to contain or to deliver a definite volume, and so forth. Proper training of personnel and continuing observation of their operations, as part of the quality assurance process, can minimize or eliminate problems associated with precise liquid measurements.

"Class A" commercial glassware, marked with a large "A," meets U.S. Federal specifications. For nearly all purposes, the calibrations of Class A volumetric glassware can be accepted, and the glassware can be used without recalibration.

The Association of Official Analytical Chemists in its *Official Methods of Analysis,* requires that burets, volumetric flasks, and pipets conform to the following U.S. Federal specifications, which are available from General Services Administration, Specification Activity 3F1, Washington Navy Yard, Building 197, Washington, DC 20407:

Item	Specification	Date
Buret	NNN–F–00780a	May 19, 1965
Volumetric flask	NNN–F–289d	Feb. 7, 1977
Volumetric pipet	NNN–P–395c	Feb. 24, 1978
Measuring pipet	NNN–P–350c	July 16, 1973

A NIST circular on this subject is available from: U.S. National Bureau of Standards Circular 602, "Testing of Glass Volumetric Apparatus." It can be ordered as Com 73–10504 from National Technical Information Service, Springfield, VA 22151.

Cleaning Glassware and Other Laboratory Ware

Clean glassware and plasticware, such as polyethylene, polypropylene and polytetrafluoroethylene (PTFE or Teflon) ware, is an essential part of laboratory operations and a vital element of the quality assurance program. Attention to the cleanliness of these items must increase in proportion to the sensitivity of the test to be made and the required accuracy of measurement.

"Clean" is a relative term. Each laboratory must establish sound cleaning procedures for glassware and plasticware used in various types of determinations, and some of these may need to be tested to demonstrate their adequacy. In general, volumetric glassware is considered to be clean when it maintains a continuous film of distilled water over its entire inner surface, that is, will drain evenly. Glassware for use in trace metal analysis may need to be cleaned with 30% nitric acid, followed by extended rinsing with redistilled water, to be suitable.

Ideally, general cleaning should begin immediately after the container or apparatus is used, and specific cleaning, when necessary, immediately before the apparatus is to be used again. For careful work, items with etched, broken, or otherwise damaged surfaces should be discarded. Manual or automatic washing equipment may be used with suitable detergents that can be rinsed away satisfactorily without contributing critical contaminants. In some cases, organic residues may have to be removed by treatment with chromic acid. After cleaning, the apparatus should be dried and stored under conditions that will not allow it to become contaminated with dust or other environmental substances.

Cleaning, Drying, and Sterilizing Bacteriological Glassware

Bacteriological glassware may be cleaned by mechanical or manual washing with the last rinse done with distilled or demineralized water (13). It is recommended that pipets be tested from time to time with bromthymol blue to make certain that excessive residual acid or alkali is absent. Petri dishes and dilution bottles require checking to ensure that they are free of bacteriostatic detergent residues. Drying of glassware may be at ambient temperature or in hot-air cabinets.

Glassware is sterilized in autoclaves with evenly distributed dry heat with the temperature of the load being not less than 170°C for at least 1 hour. Sterilization charts should accompany each sterilization load, and should be maintained as historical laboratory quality assurance records. Sterilized glassware must be stored properly to avoid subsequent contamination.

Recommendations

As part of equipment and supply management efforts of the quality assurance program, the following actions are suggested.
1. Establish criteria and specifications for the purchase of important pieces of equipment.
2. Check and test the equipment against specifications, before placing it in use.
3. Design a preventive maintenance program for equipment following the suggestions in this chapter, and establish suitable schedules and checks to see that the program is implemented and maintained.
4. Develop and maintain a supply management program to ensure the quality of reagents used in day-to-day operations, paying particular attention to primary reference standards, working standards, and standard solutions.
5. Decide on the kinds of purified water that are necessary, and develop suitable tests and testing intervals to ensure the quality of water used in analytical work and in the final cleaning of glassware.
6. Purchase only Class A volumetric glassware and perform the calibrations and recalibrations that are necessary to achieve reliable results.
7. Establish procedures for cleaning and storing glassware with due consideration for the need for special treatment of glassware used in trace analysis and for bacteriological purposes.
8. Discard chipped or etched glassware and damaged plasticware.

References

(1) Wilcox, K.R., et al. (1977) in *Quality Assurance Practices for Health Laboratories,* S.L. Inhorn (Ed.), American Public Health Association, Washington, DC, p. 58

(2) Layloff, T.P. (1980) in *Optimizing Chemical Laboratory Performance Through the Application of Quality Assurance Principles,* F.M. Garfield et al. (Eds), Association of Official Analytical Chemists, Arlington, VA, p. 47

(3) Horwitz, W., et al. (1977) in *Quality Assurance Practices for Health Laboratories,* S.L. Inhorn (Ed.), American Public Health Association, Washington, DC, p. 575

(4) Benedetti-Pichler, A.A. (1956) *Essentials of Quantitative Analysis,* The Ronald Press Co., New York, NY, p. 269

(5) "NBS Standard Reference Materials Catalog, 1988–89" (1988) NBS Publication 160, National Institute of Standards and Technology, Gaithersburg, MD

(6) Bernstein, C.A. (1980) in *Optimizing Chemical Laboratory Performance Through the Application of Quality Assurance Principles,* F.M. Garfield et al. (Eds), Association of Official Analytical Chemists, Arlington, VA, p. 51

(7) "International Standards Organization, Guide 30" (1981) American National Standards Institute, New York, NY

(8) *Official Methods of Analysis,* 15th Ed. (1990) Association of Official Analytical Chemists, Arlington, VA, pp. 645, 1219

(9) Lam, R.V., & Isenhour, T.L. (1980) *Anal. Chem.* **52**, 1158

(10) *Reagent Chemicals: American Chemical Society Specifications for 1960* (1960) American Chemical Society, Washington, DC

(11) "Standard Specifications for Reagent Water," ASTM D–1193–74 (1974) American Society for Testing and Materials, Philadelphia, PA

(12) "Standard Methods for the Examination of Water and Waste Water" (1975) American Public Health Association, Washington, DC

(13) Olson, J.C., et al. (1977) in *Quality Assurance Practices for Health Laboratories,* S.L. Inhorn (Ed.), American Public Health Association, Washington, DC, p. 666

Chapter 5

SAMPLE AND RECORD HANDLING

Records are the means by which any organization documents its operations and activities. To ensure that record production and handling are performed appropriately, a document must be prepared that describes the required records and the proper steps that are to be taken to produce them. This record system document would also indicate how various records are to be stored, the retention periods for each kind, and the circumstances and procedures for their destruction or other disposition. As with any set of procedures, these will work best when each person understands them; therefore, they must be in written form, current, and explained carefully to all employees. Continued attention to record handling by the laboratory director, supervisors, and staff is essential for proper laboratory management.

The quality assurance program of a laboratory must include reviews of significant records to determine their acceptability, and at the same time the quality assurance program will produce important records of its own. These records should document 2 things: the quality of laboratory performance, and the effectiveness of the quality assurance program itself in giving an accurate picture of laboratory performance. Aside from their use in maintaining a level of performance that is commensurate with the laboratory's standards of quality, such records can be used as evidence to support laboratory accreditation.

Among the many records produced by the laboratory that the quality assurance program must consider are those in which the results of analysis are reported, those that document the chain of custody of a sample, personnel records, records that report the results of research, equipment and supply inventory records, and those reporting the condition of analytical instruments.

Sample Accountability

A sample is generally the starting point for analytical work in the analytical laboratory. The sample may be delivered by mail, by freight, by courier, or directly by the individual who collected it. It may arrive in any of various containers and conditions: frozen, packed in ice, or at room temperature. The package may be sealed or unsealed, and the sample itself may be spoiled or broken. The sample may or may not be accompanied by appropriate documentation that should advise the laboratory why it was collected, what analysis is desired, and under what conditions of storage it should be held. All of these circumstances and conditions must be documented upon receipt of the sample for they could have bearing upon the quality or the significance of the analytical results. Since it is important to quality work to have the sample arrive in proper condition and with meaningful documentation, procedures that keep poor sample handling and delivery to a minimum must be established, continually reviewed, and enforced.

In most organizations, specific sampling procedures are written and the sample collectors are trained in their responsibilities. Generally, the sampling action is documented, the sample is identified and in some cases placed under seal, and special packaging, shipping, and delivery instructions are followed to effect delivery to the laboratory. The documentation consists of a collection report or similarly named document that accompanies the sample as it moves through the laboratory and subsequent administrative handling. This form, usually prepared in multiple copies for distribution to various units in the organization, may be supplemented with affidavits, dealer statements, bills of lading, or other relevant information that concerns the sample, its origin, and its significance. The collection report itself will contain the following types of information: a sample number, product name and identification, reason for collection, description of the sample and of the method of collection, size of lot sampled, codes, shipment information, date sample was collected, name of collector, how it was shipped to the laboratory, and whether or not it was sealed. If the sample is sealed, the seal includes the sample number, date the seal was affixed, and the collector's signature. The seal, usually one-time-use paper, is attached to the package in such a way that it must be broken before the sample can be obtained. Examples of forms used by the U.S. Food and Drug Administration can be found in Appendix B.

The next step in the sample accountability system is receipt of the sample in the laboratory. A dependable record of sample handling is important, and in most laboratories of appreciable size, the sample is accepted by a sample custodian who documents the action by completing a sample accountability record. This document will contain the sample number, the name of the product and date received, indicate who received it, describe the method of shipment or delivery, describe the packages received and their condition, and provide space for recording various storage locations before and after analysis. Deliveries of the sample, or portions of the sample, to the analyst, and its return, will also be recorded on this form, as will a signed statement concerning the final disposition of the reserve sample. The form that is used for this purpose by the Food and Drug Administration, the FD–421 or "Sample Accountability Record," is a two-part card. One copy remains with the sample custodian and the other moves with the sample through the laboratory and is used by a supervisor for sample management purposes. A copy of this form can be found in Appendix B. Some laboratories use a Sample Index Book for sample control. The information entered in the index book is essentially the same as that described for the card.

For the most part, primary responsibility for accurate and adequate record keeping lies with the scientist who examines the sample, following the system of record keeping mandated by laboratory standard operating procedures and monitored under the quality assurance program. When the sample is assigned for analysis, the analyst will have a copy of the collection record and the sample itself. At this point the analyst must determine whether the sample is in fact the item described by the collection record. No purpose is served by an analysis if significant discrepancies exist between the sample and the physical descriptions or identification described in the records. If it is possible to resolve the discrepancies, this must be done before analysis begins.

Records of analysis take several forms, among them bound notebooks, loose-leaf ring binders, analytical worksheets, and computer-generated reports. In many laboratories, perhaps most, raw data, or primary analytical information, is recorded in bound laboratory notebooks. The notebook is then the primary link between the analytical work itself and the information furnished to the client in a final report. Requirements for the correct reporting format and proper entries into the notebook are dictated by good laboratory practices and by the nature of the reporting system developed. These must be clearly imparted to each analyst in a written laboratory procedure document.

In addition to the notebook and the final report, a diversity of record types and record media may be used in any given laboratory or for any given project (1). In some situations pertinent information may be distributed between the notebook, computer or instrument records, computer printouts, magnetic disks and tapes, and instrument logbooks. It is essential for the analyst to make certain that charts, cards, and other types of data-containing records that are not incorporated into the notebook are properly identified and interpreted. These records must be clearly referenced in the notebook.

Notebook entries should be in ink, dated, and each page signed by the analyst. Periodically, the notebook is to be reviewed by a supervisor, who signs and dates the reviewed material to indicate that supervisory control is being maintained. If the results of analyses must be forwarded for administrative review and preparation of a final report, the use of bound notebooks may require the use of an accurate method for duplication. If hand transposition of data must be performed, errors will likely occur even when great care is exercised to prevent them.

There are advantages and disadvantages to the use of ring binders to hold loose-leaf analytical sheets (2). This system allows the easy addition of new sheets and the removal of completed sheets for duplication, but this ease of removal is the fundamental weakness of the system in that the loss or disordering of reports can occur easily and is impossible to detect, making the integrity of recorded data always questionable. The system works well when repetitive analyses are conducted with preprinted data sheets.

Most regulatory agencies rely on the use of analytical worksheets, a practice that is derived from experience with criminal trial procedures. Under the Federal Rules of Criminal Procedure (Title 189 of the *United States Code*), all documents and original notes that relate to the testimony of a government witness must be made available, upon the request of a defendant, for his or her examination, and the results of scientific tests or experiments that are to be used as evidence, and that are material to the preparation of the defense, must also be made available to the defendant on request (3). Consequently, if the analytical results are in a bound notebook, the entire notebook must be submitted to the defendant or entered as an exhibit of evidence. Irrelevant information may be disclosed, and the notebook itself may be rendered unavailable to the laboratory for a considerable period of time, and possibly be lost or damaged.

One of the best ways to monitor sample accountability in the laboratory is by computer (4). Using a relatively simple program, management can guard against sample mixup by the computer generation of a label that can be affixed to the sample container, and the initial entry of pertinent sample information into the computer. The information entered at log-in becomes part of the database, which is then built up through the manual or automatic addition of sample handling information and analytical data. Worksheet pages or reports can be calculated and printed, and the database itself later queried and manipulated for various informational and reporting purposes. Regardless of the recording system used, the analytical information generally reported includes a description of the sample, subsampling procedure and sample preparation, methods used, deviations from methods, validation and recovery experiments (if performed), standards used, source of reference materials, raw data, calculations, and description of the reserve sample and how it was prepared for storage after the completion of the analysis. In addition pertinent supporting documents such as chromatograms, spectra, and other charts are suitably identified with instrument identification, operating conditions, analyst names, sample number and date. If the reserve sample is sealed, the information placed on the seal is shown in the report. The sample is then returned to the sample custodian to be stored for whatever future action may be necessary, or until the sample is destroyed. FDA uses a form FD–431, "Analyst Work Sheet," which is shown in Appendix B, and a number of preprinted worksheets to cover special or repetitive product examinations.

In addition to the technical conclusion provided by the analyst on the worksheet, it is generally advisable for the laboratory supervisor to attach a clear and succinct summary for administrative review. It will focus the attention of nontechnical reviewers on sample results that require immediate follow-up action, and it also flags samples that require no further consideration. The Food and Drug Administration supervisory chemist attaches to the front of the worksheet a Sample Summary Sheet, form FD–465, for this purpose. A copy of this form is also included in Appendix B.

If the laboratory serves clients, it is important that the way in which results are reported be agreed upon in advance. Not only the format of the report, but the units of measure and the basis upon which calculations are to be made must be understood. This will make it easier for the client to interpret the results and understand their significance in answering the particular question he or she has about the product.

The International Organization for Standardization (ISO), in Guide 25 "General Requirements for the Technical Competence of Testing Laboratories," recommends that analytical work performed by a testing laboratory be covered by a report that clearly, accurately, and unambiguously presents the test results and all relevant information (5). The Guide suggests that the test report include at least the following information:

- Name and address of the testing laboratory
- Unique identification of the report (such as series number), and each page of the report
- Name and address of the client
- Description and identification of the test item
- Date of receipt of the test item and dates of performance of the tests
- A statement to the effect that the test results relate only to the items tested
- Identification of the test specification, method, and procedure
- Description of the sampling procedure if the laboratory drew the sample
- Description of the preparation of the analytical sample
- Any deviations from, additions to, or exclusions from the test specification, and any other information relevant to a specific test
- Disclosure of any nonstandard test method or procedure used
- Measurements, examinations, and derived results, supported by tables, graphs, sketches, and photographs as appropriate, and any failures and discarded data, which should be so identified
- A statement on measurement uncertainty (where relevant)
- Signature and title of the people accepting technical responsibility for the test report and dates of issue
- A statement that the report shall not be reproduced except in full without the approval of the testing laboratory

The ISO Guide also suggests that particular care and attention be paid to the arrangement of the test report, especially with regard to presentation of the test data for ease of understanding by the reader.

Maintenance of Analytical Records

The maintenance of records of analyses is as essential to a laboratory's operation as the various steps in the collection, analysis, and storage of samples. The records have potential long-term value and may serve many purposes. If a record is lost because of carelessness, the time spent in analysis may be wasted. Sloppy records and poor records maintenance likely reflect poor quality control in other areas of operation, and will give that impression to others.

When the worksheet system is used, it is a good practice to keep all records, such as the collection report and its attachments, the analytical report and its attachments, and correspondence associated with a particular sample in a single folder, clearly labeled and filed in a prescribed manner. The file must be identified with the sample number, case number, or by some standardized means that permits ready retrieval. Because of the odd sizes of instrument charts, exhibits, and so forth, care is necessary in mounting these items and arranging them so that they are fully contained in the folder and not easily lost. It is not advisable to remove records from folders, leaving the folder and partial contents in the file; the entire folder should be available when necessary, and its location monitored with a sign-out and suspense file system to ensure that folders are returned to their proper locations. The fewer people allowed to add or remove folders or records, the better the chances are that the material will not be misfiled or lost.

When bound notebooks are used to record analytical results for samples, the laboratory has the problem of establishing a standardized notebook control procedure that covers both the issuance of notebooks and their storage when they are filled (6). It is recommended that all notebooks be numbered consecutively and a central record made that contains, for each log issued, the book number, type of log, date of issue, name of person to whom assigned, date of return, and storage location. Assignment of the responsibility for laboratory notebook control to a single monitor is quite important, but individual analysts must be responsible for the use and care of their own notebooks until they are completed and turned in. Obviously, the use of notebooks for recording analytical data can lead to inefficiencies. If a multiple notebook system is used, for example, several analysts using one notebook for foods, one for drugs, and so forth, rather than individual notebooks, other problems can arise. In either case, the procedure adopted must be documented and a periodic review of its proper implementation made a part of the quality assurance program.

Retention of Records and Samples

After an analysis is complete and the results are reported, the laboratory needs a written policy for guidance on the retention of samples and the associated records. For samples and records that may be involved in litigation, the storage period can extend for years. For the majority of samples, fortunately, this is not usually the case. The object should be to destroy samples as soon as it can be determined, with certainty, that they will no longer be required for further testing or as evidence, and to dispose of records after they are no longer legally or administratively important. Storage periods, obviously, must be determined by each facility, depending on its obligations, but the point to be made here is that a clear policy must be in place to prevent both the destruction of important items and the accumulation of what is essentially junk. From a quality assurance point of view, the improper destruction of active samples or records is low quality performance in violation of policy, and the QA program must provide a means to detect such actions in an effort to prevent their recurrence.

Computerized Records

Although a great deal of data are produced in the laboratory, to be useful it must be processed into information. Computers can be used to good advantage to accept the data, process it, and produce a report that contains the resultant information (7). When handled properly, computers can assist in producing more timely and more accurate information than can a manual system.

Two types of laboratory information systems have to be distinguished — one that deals with management of the workload, and one that deals with the workload itself (the analyses). Actually there is a third information system — one that does both. A workload management system will tell the supervisor and manager such things as what samples have been received, where they have been assigned, how long they have been in the laboratory, when they were completed, who analyzed them, how many hours the analyst required, and whether or not the sample was within specifications. A computerized analytical system, on the other hand, takes its information from analysts at keyboards or directly from analytical instruments and produces worksheets, or analytical reports those contain data and calculations derived from those data. Either of these systems can be designed to generate summary or other special reports, and to send reports to terminals or printers at various locations.

Computerizing a laboratory can be quite difficult, in spite of the ease claimed in promotions by computer hardware and software vendors. It may not be desirable from a management standpoint, in fact, to reorganize a small laboratory into a computerized operation. The effort, however, can be not only worthwhile but a competitive necessity for a larger laboratory, where numerous samples are to be processed, or many people or locations need the information that is produced, or where there is a need to collate and disseminate information quickly.

A type of laboratory computer system that has received widespread acceptance is the laboratory information management system (LIMS). LIMS can be defined as a database tailored to the analytical laboratory so that it can handle data generated by the analysis of samples and integrate sample information with results obtained from analytical instruments, thereby reducing administrative tasks and increasing the production of final reports. The LIMS is the system mentioned above "that does both." It provides tracking, database query, integrated graphics, data archiving, audit trails, and report formatting.

There are essentially three ways to obtain a suitable LIMS: The system can be developed in-house, it can be developed by an outside expert under contract, or it can be purchased as a package from a vendor. There are advantages and disadvantages to each procedure, and the advantages and disadvantages are not the same for all organizations.

In-house development is excellent for a large corporation with an experienced software development department provided with enough money and employees to complete an assignment such as this in a reasonable time. The main advantage, of course, is that of designing exactly the program that is needed for a specific organization. In addition, the user will know enough about that program to be able to make changes in it as change is deemed appropriate, and will not have to share the program or pay any license fees. The usual disadvantage is not being a large corporation. Striking out on one's own using the overtaxed resources of the local computer hacker will normally produce a great deal of frustration, eventually a serviceable but inferior program, and the loss of the productive services of the person developing the program — not to mention those who try to use it.

Outside LIMS development is preferable for a laboratory that wishes to develop its own program but does not have the internal resources to do it properly. A better program can be obtained than one can develop oneself with limited amateur resources, and the program will be tailored to the laboratory's needs to the degree that the laboratory staff can make itself clearly understood by the consultant. This procedure can be expensive, and, in comparison to the third choice, slow to install and to make operate efficiently ("debug").

Commercially purchased programs do not allow the control by the purchaser that the previous 2 options do, but there are a fair number of package systems available from among which one can probably make a reasonable choice. The program will probably not need debugging, but the fit will not be perfect, and the laboratory may have to live with a less than perfect system.

If a decision is made to acquire a computer system, then everything possible must be done to get the best possible system. It is certainly advisable that a computer system quality assurance team be appointed. The purpose of this team would be to describe every important operation of the laboratory in simple, specific terms, indicate priorities and time requirements, decide which activities should be computerized, establish short- and long-term objectives that computerization must meet. Verification requirements, laboratory program needs, and quality assurance checking procedures must be developed into the system. The team must take into account the experience of the people entering the data or obtaining reports and consider human engineering in the system design that will minimize errors or distortion of the data. Procedures for system validation and future revision also must be developed. Finally, consideration needs to be given to the control of source documents, the inclusion of suitable audit trails, personnel training requirements, data security, and appropriate backup systems that will allow recovery of data in the event of system failure.

Quality must be built into the system. Murine (8) suggests a number of conditions that serve to define quality software.

- Correctness: the extent to which a program satisfies expectations and fulfills objectives
- Reliability: the performance of functions with the required precision
- Efficiency: the use of minimum resources to perform the functions
- Integrity: providing control of access to data — security
- Useability: ease of effort required to use the system and to interpret the output — "user friendliness"
- Maintainability: the ease locating and correcting errors and debugging the program
- Testability: ease of testing the software to ensure the proper performance of intended functions

There are costs involved in producing a quality computer system. The majority of the costs are the preparatory costs of establishing the design criteria, and then developing or purchasing the software. The equipment that can run the software may also represent a significant cost. There are also costs for quality assurance since significant time must be devoted to preventing errors and to appraising and correcting system performance. Estimates for the cost of quality assurance in the development of a computer software quality system range from 5 to 50% of the total development cost, a generally accepted figure is 9% (8).

A computer LIMS can be used to advantage in many ways in a laboratory that will serve to ensure quality. Some of these are sample accountability; analytical methods storage and retrieval; preparation of sample reports; maintenance of inventories of equipment, reagents, and supplies; evaluating intra- and interlaboratory test sample results; preparation of analytical control charts; maintaining various personnel and training records; monitoring laboratory research; and providing financial planning and cost accounting information.

The computer has already become, like the photocopier, something we cannot imagine being without. The challenge is to find how it can be used to improve a specific operation, such as that of the small laboratory, without sacrificing the precision, accuracy, security, and convenience that the operation requires. New software programs continue to be developed, the development often based upon new insights into how a set of activities can be analyzed and then duplicated in computer program format. Although the management of a laboratory may not choose, now, to add the computer to its equipment inventory and a computer program to its management tools, they would be wise to make themselves aware of the possible advantages, and be ready.

Recommendations

In establishing sample and record-handling procedures, consider the following:
1. A suitable sample collection report form for documenting sample collection and delivery of samples to the laboratory.
2. A record system for receipt of samples in the laboratory, delivery of samples to analysts, and storage of samples after examination.
3. Documentation of analyst responsibilities in sample handling, analysis, and reporting.
4. A system for reporting analytical results to clients or for use within the organization.
5. A filing system for worksheets, collection records, notebooks, and so forth.
6. A policy for the disposition of documents and samples when they are no longer needed.
7. Use of a computer system for sample handling and analysis, and for various management purposes, bearing in mind the design of the system for quality performance, costs, and procedures to ensure that the data entered into the system are accurate and readily retrievable.

References

(1) *Quality Assurance Handbook—Center for Analytical Chemistry* (1987) National Institute of Standards and Technology, Gaithersburg, MD, p. V–1

(2) McCully, K.A., & Lee, J.G. (1980) in *Optimizing Chemical Laboratory Performance Through the Application of Quality Assurance Principles,* F.M. Garfield et al. (Eds), Association of Official Analytical Chemists, Arlington, VA, p. 80

(3) Frank, R.S. (1980) in *Optimizing Chemical Laboratory Performance Through the Application of Quality Assurance Principles,* F.M. Garfield et al. (Eds), Association of Official Analytical Chemists, Arlington, VA, p. 136

(4) Dux, J.P. (1986) *Handbook of Quality Assurance for the Analytical Chemistry Laboratory,* Van Nostrand Reinhold Co., New York, NY, p. 69

(5) "International Standards Organization, Guide 25" (1982) American National Standards Institute, New York, NY

(6) Dux, J.P. (1986) *Handbook of Quality Assurance for the Analytical Chemistry Laboratory,* Van Nostrand Reinhold Co., New York, NY, p. 60

(7) Siggins, R.C. (1989) "Records and Reporting Laboratory Information Management Systems," presented at an AOAC Quality Assurance Short Course, September 28-29, 1989, St. Louis, MO, Association of Official Analytical Chemists, Arlington, VA

(8) Murine, G.E. (Nov. 1988) "Integrating Software Quality Metrics with Software QA," *Quality Progress*

Chapter 6

SAMPLING AND SAMPLE ANALYSIS

There are three basic activities involved in solving an analytical problem:
- Collection of a representative sample
- Preparation of the sample for analysis
- Analysis using appropriate methods

These activities are independent of each other, yet one can have decisive influence on another. Since there is a potential for error in each of these activities, steps must be taken to identify these errors and avoid them. Laboratories should develop a plan for the proper performance of each activity, then establish quality standards and written procedures that will meet the standards. The development of appropriate plans will depend upon an understanding of the problems involved in each activity, and then the application of reasonable judgments in seeking solutions.

Sample Collection

The laboratory staff is usually not involved in planning the collection of samples, and this is unfortunate. Although analysts may be consulted from time to time concerning proper sample size or the use of appropriate sample preservatives, and asked to provide suitably prepared containers, often even this may not occur, and the analyst is left to make the best of a poor situation. This is unfortunate for the laboratory because, when disparate analytical results are obtained from different samples of the same lot, the usual conclusion is to attribute the difference to poor analytical work or to a poor choice of methods rather than to the more likely cause, which is poor sampling of the lot of material. When a significant difference in results occurs between laboratories that supposedly have analyzed the same sample, a serious problem may arise involving questions of competence and credibility. Many of these situations can be avoided if samples are collected according to a rational plan that gives some assurance that the sample delivered to the laboratory represents the composition of the parent lot.

There are at least two ways to measure a given lot of goods: One, often assumed to be the "proper" way, is to find its "true value," by which one means its *average* value. Another, often discovered accidentally as the result of "poor" sampling, is to measure its variability. So called proper sampling of drug dosage forms, for example, may involve compositing 20 tablets, by which process the majority of the tablets could be used to dilute and conceal the fact that several of them are severely sub- or superpotent. Similarly, two lots of grain may have been purposely, but ineffectively, mixed in an attempt to reduce the average level of a contaminant. Sampling that led to the laboratory finding of inconsistent results would reveal the attempt to dilute an illegal product.

Few studies have been conducted on the distribution of error among the three activities: sampling, sample preparation for analysis, and analysis. In one such study, which involved 20 parts per billion of aflatoxin in a peanut sample, the error contributed by the original sampling of the lot was 67% of the total standard deviation, the error contributed by the analyst preparing a subsample was 20%, and the analytical procedure error was 13% (1). This example exaggerates the proportion of error that can be attributed on the average to sampling, because the distribution of aflatoxin in peanuts can vary widely, with a few peanuts accounting for most of the contamination. The important point in this example, however, is to show that sampling error can play a very significant part in the overall error in the analytical system.

The Sampling Plan

Sampling is generally done for a specific purpose and the purpose may, indeed, suggest or dictate the nature of the sampling plan. The presence of a well-designed plan is important since it provides a consistent model to guide the people performing the sampling activity, and it serves as a reminder of the important elements in this part of the overall sample analysis program. The sampling plan serves several additional purposes: it requires the thorough consideration of collectors and analysts before it is issued — a meeting of the minds of the persons concerned; it serves as a reference source; it provides the means for operating on a planned basis; it can serve as a source for training; and, very important, it furnishes a means for comparison of performance against objectives.

In many sampling programs, statistical sampling approaches are not always given the attention they deserve. The old and often-used square-root sampling method is being discarded, as it should be, because of its questionable value. Percentage sampling systems that specify a fixed percentage of a lot, say 5% or 10%, do not provide the quality protection that is often assumed. Statistical sampling theory furnishes the means to analyze the relationship between a lot of goods and the samples that are drawn from it. It can be used to estimate population measures, or "parameters," such as variance and correlation, from a knowledge of corresponding sample quantities.

There are several sample selection methods in common use: probability sampling, nonprobability sampling, bulk sampling, and acceptance sampling. These are described to a limited degree below:

Probability sampling is used when a representative sample is desired, and involves principles of statistical sampling and probability, a random selection approach that tends to give each unit an equal chance of being selected.

Nonprobability sampling is used when it is not possible to collect a representative sample, or a representative sample is not desired. The sample collector uses judgment rather than statistical considerations in the selection of the sample.

Bulk sampling involves the selection of a sample from a lot of material that does not consist of discrete, identifiable, or constant units. Sampling may be performed in static or dynamic situations. Bulk sampling poses special problems requiring certain decisions to be made: the number of increments to be taken, the size of the increments, from where in the pile or stream they should be drawn, the sampling device to be used, and how to reduce the increments taken to a reasonable size sample for delivery to the laboratory.

Acceptance sampling differs from the previous examples, which can be characterized as *survey sampling*. Acceptance sampling involves the application of a predetermined plan to decide whether a lot of goods meets defined criteria for acceptance. The risks of accepting "bad" or rejecting "good" lots are stated in conjunction with one or more parameters, for example, quality indices of the plan. Statistical plans can be designed to regulate the probabilities of rejecting good lots or accepting bad lots.

There are two broad categories of acceptance sampling. These are called sampling by attributes (*2*) and sampling by variables (*3*). In *sampling by attributes* the unit of product is classified as defective or nondefective, or the number of defects in the unit of product is counted with respect to a given requirement. In *sampling by variables* a specified quantity characteristic or a unit of product is measured on a continuous scale (for example, pounds, inches, feet per second) and a measure is recorded.

An example of net weight determination may serve to explain the differences between the two categories (*4*). In attribute sampling, each unit that weighs 1 pound or more is accepted, and each unit that weighs less than 1 pound is rejected. If the number of rejects exceeds a predetermined number, the lot is rejected. If the number of rejects is less than the predetermined number, the lot is accepted.

In variable sampling, if the weights are averaged and the average meets or exceeds the declared weight with no reasonable shortages as judged by the standard deviation, the lot is accepted.

Operating characteristic (OC) curves are used extensively in acceptance sampling. The OC curve shows the relationship between the quality and the percent of lots expected to be acceptable for the quality characteristic inspected. In other words, the OC curve is a graph of lot defectives against the probability that the sampling plan will accept the lot. Figure 1 depicts OC curves for an ideal sampling plan and an actual one.

United States Military Standard 105D (*2*) includes a variety of tables and OC curves. It provides seven inspection levels that cover varying levels of discrimination, that is, "tightness" or "steepness" of OC curves, and three levels of inspection in terms of severity of inspection, that is, "normal," "tightened," and "reduced."

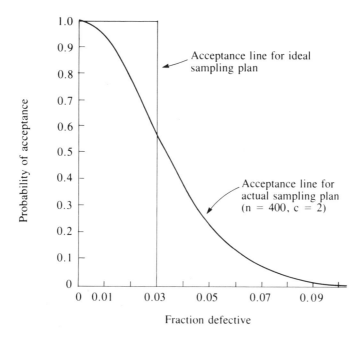

Figure 1 OC Curves for Ideal and Actual Sampling Plans

MIL–STD–414 (3) is a companion publication for examining lots by variables. It presents curves for a variety of plans, but the plans do not match those in MIL–STD–105D.

Four excellent articles on sampling are listed under *References* and should be consulted for additional information on this topic (5).

Samples are useful for their intended purpose when they are taken in a manner consistent with generally recognized good sampling techniques and practices. This requires inspection of the lot before sampling; use of suitable sampling devices for the particular commodity and type of sample desired; use of suitable containers to hold the sample; maintenance of the integrity of the sample and associated records; use of adequate precautions in preserving, packing, and delivery of the sample to the laboratory in a timely manner; and provision of appropriate storage conditions for the sample both prior to and following analysis. All of these factors, along with others such as a cost-benefit analysis and a review of program objectives and regulatory requirements, are to be assessed and brought together in a firm sampling plan that serves as a guide to management as well as to operating personnel in achieving quality in sampling.

The development of quality sampling plans is a science in itself and has been given consideration by a number of organizations. One plan format that deserves serious consideration, developed by the International Organization for Standardization, is shown below, with comment (6). It can serve as a starting point, or checklist, for developing a sampling plan for most commodities. The title and headings from sections in the monograph are as follows:

"Agricultural Food Products — Layout for a Standard Method of Sampling from a Lot"

1. Title (short but appropriate for index identification)
2. Introduction (describing the purpose of the plan)
3. Scope (describing the breadth of coverage of the plan)
4. Field of Application (products to be covered; where sampling will be done)
5. References (documents the validity of the plan with reference to other requirements)
6. Definitions
7. Principle (statistical basis of the method of sampling)
8. Administrative arrangements
    8.1 Sampling personnel
    8.2 Representation of parties concerned
    8.3 Health, safety, and security precautions
    8.4 Preparation of the sampling report
9. Identification and inspection of the lot prior to sampling (important in survey sampling for identification, condition of the lot, and selection of the method of sampling)
10. Sampling equipment and ambient conditions (proper tools, such as use of sterile equipment for aseptic sampling)
11. Sample containers and packing (essential to prevent contamination and damage during shipment or storage)
12. Sampling procedures (as dictated by the plan objectives)
    12.1 Sample size
    12.2 Taking of the sample
    12.3 Preparation of bulk samples and reduced samples
    12.4 Selection of samples of prepackaged products
13. Packing, sealing, and marking of samples and sample containers (identification of units and to establish chain of custody)
    13.1 Filling and sealing of sample containers
    13.2 Marking
    13.3 Packing samples for storage or transportation
14. Precautions during storage and transportation of samples
15. Sampling report
    15.1 Administrative details
    15.2 Details of unit packs or enclosure containing the lot
    15.3 Material samples
    15.4 Marking and sealing of samples
16. Annexes (supplemental information, if necessary)

Subsampling for Analysis

The best analysis can give misleading information if the sample portion analyzed does not represent the sample or the lot from which it was taken. Distortions introduced at this point will carry through the analysis and adversely affect the final results and the conclusions drawn from them. There are generally two choices in analytical subsampling: preparation of a composite laboratory sample (if multiple units are submitted for analysis) or examination of individual units. A *composite* laboratory sample is one in which the individual units, or representative portions of units, are mixed to form a uniform mixture. Portions are then taken from the composite for analysis. Compositing can best be used when homogeneity is not a significant concern. Compositing saves analytical time and, in some types of contract testing, it may be the procedure specified. It is not the procedure of choice when there is a chance that an individual unit that constitutes a public health or safety threat will not be detected (there are some exceptions) or where a unit at or outside of tolerance will not be detected because of matrix dilution. *Multiple unit* laboratory sampling is indicated when the possible range of values among individual units is considered significant or it is desirable to establish the variability of the lot.

Youden and Steiner (7) observe that "Many materials are notoriously difficult to sample. Often the variability among samples is the controlling factor in the confidence placed in the analytical result." They note further that

> A mistake sometimes made is to composite several samples and then to run repeat determinations on this composite sample. The analyst may be happy with several results that are in close agreement because only the analytical error is involved in the results. And some may put their faith in the result. Admittedly, if the individual samples were of the same weight and properly mixed, the same average will result whether the samples are analyzed individually or repeats are made on the composition. Using the composite sample effectively conceals the between-sample variation. It should be mandatory to run the samples individually, for only by doing so will anybody be in a position to make any statistical statement about the results, no matter how good the analytical procedure.

A somewhat similar view of subsampling for analysis is expressed in an article published in *Chemical and Engineering News* by an ad hoc Subcommittee of the American Chemical Society for "Dealing with the Scientific Aspects of Regulatory Measurements" (8). This report observes that

> The number of samples to be analyzed in a given situation usually is limited by the resources available for the collection of the samples or for their analysis. However, the reliability of the result generally increases with the square root of the number of samples analyzed. For this reason, analyses of multiple samples always are preferred over single samples since single samples give no information on the homogeneity of the lot that was sampled. In addition, for single samples, the sampling error is also confounded with the analytical

error. As a result, if the total number of determinations must be fixed, multiple independent single samples are preferred over replicate aliquots from a single sample. If only a single analysis is possible, a composite sample is preferred over a single random sample. In any case, the sampling decision should be an *a priori* decision and should be based on the question at issue.

In addition to the number of subsamples taken for analysis, it is essential that each be prepared in a way that achieves homogeneity and is handled in a manner that prevents alteration from the original composition. Obviously, failure to prepare a homogeneous subsample at this point will affect the results of the analysis regardless of the method used.

Sample Preparation for Analysis

Every type of material that is to be prepared for analysis presents its own practical difficulties. The requirements for suitable sample preparation are dictated by the consistency and the chemical characteristics of the analyte and the matrix, and by the distribution of the analyte in the sample. Even seemingly homogeneous materials such as liquids may be subject to sedimentation or stratification. Vigilance and care are the watch words.

Single-phase liquids can generally be mixed, stirred, shaken, or blended. Dry particulate materials can be reduced in volume by coning and quartering, by rolling and quartering, or by the use of a splitter, such as a riffle (9). A variety of implements and machines are available for sample disintegration, such as mills, grinders, and cutters. Care in their use is necessary to prevent loss of dust or change in composition through the partial separation of components. Screening can be used to improve the efficiency of particle size reduction, followed by mixing to attain homogeneity. Sampling errors can occur even in well-mixed particulate mixtures, especially in trace analyses, if the particles differ appreciably in size or physical properties.

Every piece of equipment used in the preparation of a sample must be viewed with suspicion. Grinders were mentioned above as contributing to the loss of finer particles as dust. They have been known to segregate materials within the mix by size as well, with the finer material collecting beneath the blade, for example. Metal screens can pass fine particles, but retain powder that adheres to the screen material. Glass containers and laboratory apparatus can adsorb certain materials and may require surface treatment. Plastic containers can hold contaminants, such as animal hairs, while the rest of the sample is transferred with apparent ease. In other words, validation of a method of analysis, to be discussed below, most certainly includes validation of the method of sample preparation and storage.

Loss or gain of moisture during manipulation can be a problem. Loss can be minimized by keeping samples covered with plastic or aluminum foil. A cold product can be protected from gaining moisture by allowing the sample to come to room temperature before preparation begins.

When volatile organic constituents are present in samples, sample manipulation may not be possible, or may be severely restricted, in order to prevent their loss. This complicates sample preparation.

As a general guide, food samples are analyzed in the form in which they are commonly consumed. Inedible portions, such as peel, nut shells, or fish bones, are removed and discarded prior to analysis, and suitable note made of how the sample was prepared.

Trace metal analyses can present significant problems (*10*). For example, the trace metals can be distributed unequally between liquid and solid phases in canned vegetables and canned fruits. Most people do not consume the vegetable brine, but the liquid portion of canned fruits is generally eaten. Obviously, this irregular distribution of metals can pose problems for the analyst in establishing the level of the metal residues in the product, as well as for those concerned with setting tolerances.

As mentioned earlier, one of the most difficult problems in sampling from a lot, and in subsequent laboratory subsampling, is encountered in trying to obtain a representative sample for the analysis of aflatoxins in raw agricultural commodities (*11*). Aflatoxin contamination exhibits a highly erratic distribution, with a reduction in heterogeneity as the food or feed is reduced in particle size. After it was recognized that there was a high rate of variability between and within samples from the same lot, there was a movement toward the collection of larger and larger samples. Sample sizes started, for peanuts, with 1 kilogram, and the size increased as more reliable results were required by food producers, since increasing sample size reduces the number of good lots that are likely to be rejected and the number of bad lots accepted.

At the present time in the United States, the sample taken from a lot of shelled peanuts is 144 pounds — three 48-pound samples — with portions taken at random from the lot. Examination in the laboratory is by sequential analysis with the first 48-pound sample ground in a subsampling mill and test portions examined in duplicate. If the average of the test portions is below the established tolerance (set by the Food and Drug Administration) the lot is passed. If the average is above the acceptance level, the lot is rejected. If the findings fall between the two figures, the second 48-pound sample is comminuted and the analysis repeated. If a decision cannot be made to accept or reject the lot, the third 48-pound sample is prepared, assayed, and the cumulative results considered. The foregoing example points out dramatically the need for attention to lot sampling, laboratory subsampling, and sample preparation for analysis. Although this is a rather extreme case, it illustrates that sampling problems cannot be ignored or treated indifferently.

Method Selection

Analyses are intended to provide reliable information on the nature and composition of materials submitted for analysis. It is well-recognized that a certain degree of variability is associated with all measurements. This variability usually increases when the measurements are made by different analysts in the same laboratory, and even greater variability is observed when the analysts are in different laboratories (*12*).

One objective of a quality assurance program is to hold this variability in analytical findings to a minimum.

Methods of analysis have certain attributes such as accuracy, precision, specificity, sensitivity, dependability, and practicality that must be considered when choosing the most appropriate method to solve a specific problem. It is not always possible or even desirable to optimize all of these attributes during any given analysis. Ultimately the analyst must evaluate all of the information at hand and decide on the level of uncertainty that is acceptable in a particular situation. This scientific information must be balanced against practical considerations such as analysis time, cost per analysis, error risk, and level of expertise required of the analyst for satisfactory performance. These attributes are sometimes, referred to as "figures of merit" (*13*).

As a minimum the performance data provided for a method should include the following information (*14*): evidence of analyte identification; evidence of separation of the analyte from interfering substances; the lower level of measurement of analyte concentration; a reasonable measure of precision attainable within the laboratory, and, if possible, between laboratories, at functional levels of the analyte in the matrix; and the accuracy of measurement of the analyte at typical levels of concentration. In other words, specificity, sensitivity, reproducibility, and accuracy are required.

The precision with which an analyte can be determined depends on the analyst, laboratory conditions, the concentration of the analyte, types and nature of the interferences or contaminants, limit of detection, and the integrity and stability of the

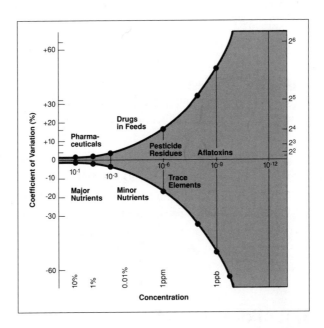

Figure 2 The general curve relating between-laboratory coefficients of variation (expressed as percent on the left and powers of 2 on the right) with concentration (expressed as powers of 10 along the horizontal center axis) *decreasing* toward the right. A convenient reference point is at 1 ppm (10^{-6}) where CV = 2^4 = 16%

analyte. These influences affect reproducibility more significantly as the concentration of the analyte decreases. Horwitz (*15*) studied this effect by the examination of the cumulative results of AOAC collaborative studies. When he plotted the coefficient of variation as a function of concentration (see Figure 2) expressed in powers of 10, he found that the coefficient of variation doubled for every two orders of magnitude *decrease* in concentration. At a level of 1% or more, the coefficient of variation between laboratories was about 2.5% and at one part per thousand it was 5%. At the parts-per-million level the CV was 16%, and at the parts-per-billion level it was 32%. The interesting feature of the curve is that it appears to be independent of analyte, matrix, method of analysis, and measurement technique.

The selection of appropriate methods often must take into account matters other than those attributes dealt with above. Methods can be loosely classified according to their purpose or their administrative propriety.

- Official methods
- Standard Consensus, or reference methods
- Screening, or rapid methods
- Routine methods
- Automated methods
- Modified methods

Official methods are generally those required by law or regulation for use by a government agency or an industry regulated by a government agency. For example, the Food, Drug, and Cosmetic Act, under which the Food and Drug Administration operates, cites the methods in the *United States Pharmacopeia* and requires their use for the analysis of items described in that compendium. *Official Methods of Analysis of the AOAC* is designated by FDA regulation as methods the agency uses in regulatory analyses of other products for which methods are provided in that publication. *Standard consensus,* or *reference methods,* are those developed by organizations or groups that use collaborative studies or similar approaches to validate them. Their value depends upon the authority of the organizations that sponsor them. *Screening* or *rapid methods* are used as an expedient means to determine, for a large number of samples, whether any of them should be subjected to additional testing by a more accurate method. *Routine methods* are those used on a routine basis. They can be official or standard, but may be modified to be more convenient to use on large numbers of similar samples. *Automated methods,* as the name implies, use automated equipment but may also be official, or screening, methods. *Modified methods* are generally official or standard methods that have been modified to simplify them, or to accommodate them to different types of samples than those originally envisioned, or to remove uncommon interfering substances.

Proper method selection is critical to laboratory performance and must be monitored under the quality assurance program, but there is, in fact, no simple, established formula for it. Cobb (*16*) suggests that a combination of efforts can provide a reasonably good fix. The assessment process may entail a literature review, reference to personal experience or the experience of colleagues, consideration of what instruments are available, the amount of time available for analysis, the accuracy and precision required, and similar considerations. The National Institute of Occupational Safety and Health has developed an empirical system for circumstances where there is no legal requirement to use a specific method (*17*):

- Methods that have been applied to the matrix of interest should be preferred over methods that have been applied to other matrices or methods that apparently have not been tested on authentic samples.
- Methods documented by published interlaboratory validation data should be selected over those that are not.
- Methods that have been tested and validated over the concentration range of interest should be chosen over methods tested at other levels (particularly higher levels). Methods that perform quite well at one level may be totally inadequate at a lower level.
- Methods that are widely used should be chosen over methods not in wide use.
- Methods that are simple, low cost, or rapid should be chosen over methods that are complex, more costly, or slower.

Egan (*18*), using similar considerations, suggests that

- Preference be given to methods for which reliability has been established in collaborative studies in several laboratories.
- Preference be given to methods that have been recommended or adopted by relevant international organizations.
- Preference be given to those methods of analysis that are uniformly applicable to various substrates over those that apply to individual substrates.

Even when a method is selected and found to be suitable for a given type of sample, control checks should be conducted along with the analyses, or intermittently after a specified number of analyses, to document its consistent performance. It is also useful to construct control charts as described in Chapter 2, using results from control samples or selected function checks to verify the consistent quality of performance. Control charts provide a ready check on whether or not the analytical system is in or out of control and can indicate which part may be the probable cause.

Method Validation

Method validation is a process by which the attributes or figures of merit are determined and evaluated, and is an important part of the quality assurance program (*19*). Many methods are validated by consensus standards organizations. Some methods must be used because they are official, as mentioned earlier, regardless of how or whether they have been validated. As a general rule, official methods that are applied to the type of sample they were specified for, and methods that have been collaboratively studied on the matrix and concentration of interest, need not be validated before use. On the other hand, the use of those methods will not always ensure desirable results. An analyst who has some experience in the use of the method will have a significantly increased chance of a satisfactory analysis, but it is also important to remember that a method that is valid in one situation may not be valid in another.

So if the analyst is inexperienced with the method, or the analytical situation is unusual, revalidation is the safer course.

Several approaches are available for method validation, and they should be given consideration especially if a method has not been used by the laboratory or the analyst before, or if it is is one that will be used by the laboratory in the future on a more-or-less regular basis (20). These approaches are:

- The analysis of synthetic formulations
- The analysis of spiked samples (standard addition)
- Comparison of results with those obtained using an official or standard method
- The analysis of certified reference materials of known concentration

The ideal approach to method evaluation is to test a sample of known composition that exactly duplicates the type of sample on which the method will be used (analyte of interest, proper concentration range, and identical matrix). Unfortunately, this ideal solution is not always feasible. Perhaps the total composition of the sample is unknown, or the authentic ingredients are not available. An even more complicating factor is that the manufacture of the product, or biological inclusion of a residue, may incorporate the analyte in a form that cannot be duplicated readily by the analyst. It is therefore often necessary to resort to other approaches of method validation.

Standard addition procedures involve the addition of known quantities of pure substance to portions of previously analyzed sample, and repeating the analysis using the same reagents, instrument, and technique. Two procedures are described here. In the first, the sample of interest is analyzed in duplicate to ensure that results from various portions of the sample composite will be consistent, that is, the method will give a reasonable precision. Then, a half portion of the sample composite is taken and spiked with an accurately measured amount of the analyte equivalent to about half of that found in the original analysis. The percent recovery for the method is then calculated using the ratio of the amount found (F), less the amount already present [half of that found in the original analysis (O)—equivalent to a blank] over the amount spiked or added (S):

$$\% \text{ Recovery} = \frac{F - \frac{1}{2}O}{S} \times 100$$

In the other procedure, portions of sample are spiked at levels that bracket the estimated analyte concentration in the test sample, and a calibration curve is constructed. If the calculated concentrations after analysis are a linear function of the added spikes, performance can be considered to be satisfactory. If the line passes through the origin on the graph, and the unspiked test sample shows no response for the analyte, it is reasonable to assume an absence of analyte in the sample. If a response is shown by the unfortified sample and the response lies on the same line as the spiked samples, the line is extrapolated into the negative x-axis region. The quantity of analyte in the unfortified sample is reflected by the point where the negative x-axis is crossed, but read on a positive scale (the mirror image of the normal concentration scale on the positive x-axis) (16).

The standard addition procedure is not as rigorous as using a good synthetic formulation. The method is especially useful in situations where there is a constant background of interference from matrix elements and it is not possible to suppress them.

It is important that the standard addition or spike be made early in the analysis, either to the original test sample portion taken or soon after, and not in the last step. If addition is made in the last step, the validation only extends to the addition of the spike and the subsequent measurement step. This is another common disadvantage to the meaningfulness of standard addition recoveries as compared to the use of a standard sample, because it is frequently easier to recover the pure substance that is added to a prepared extract than it is to extract the compound from the original sample material. In other words, adding the spike at a late stage in the analysis may give false high recoveries.

The other two validation procedures are self-explanatory. Comparison with an official method, or with a method that has been found to be acceptable, can be used if the samples used in the comparison and the sample to be tested by the newly validated method do not differ substantially in composition or in the order of magnitude of the analyte concentration. Finally, certified reference materials, such as those available from the National Institute for Standards and Technology, can be used to calibrate methods that are to be used to detect the specific analytes in similar products.

Ruggedness Testing of Methods

Some methods are sensitive to relatively minor changes in analytical technique, reagents, or environmental factors. Youden and Steiner suggest the use of "ruggedness" testing as protection against such sources of trouble. The plan is to introduce minor reasonable variations in the method to see what happens. Their plan introduces several changes at once in such a manner that the effects of individual changes can be ascertained using a limited and reasonable number of analyses (21). The standard deviations calculated from the limited number of runs provide a good measurement of the level of imprecision that can be expected in the routine use of the method in the laboratory. It is suggested that all new analytical methods, developed in or used by the laboratory on a regular basis, be subjected to ruggedness testing with the aim to design and adopt only methods that are suitably stable.

Methods Control

Once methods have been adopted for use in a laboratory they must be exactly followed by analysts, without significant deviations. The permissible type and range of modifications must be clearly understood, and all such changes thoroughly documented in the original analytical record. More substantial changes, when they are not accompanied by validating recovery experiments, can and do create problems. In situations of this type, poor results are generally caused by the questionable performance of the analyst rather than by the performance of the method.

Dux (*22*) suggests, for quality assurance purposes, documenting the authorized use of methods by the laboratory director. He recommends the use of a "Method Authorization Form" to prevent the use of untested or unproven methods. The form includes title of the method; analyte; matrix; brief description of the scientific basis of the method; a brief description of the results of the validation study that covers accuracy, precision, interferences, applicable concentration ranges, limit of detection, result of ruggedness test, name of the analyst who performed the study, location of the data (notebook number and pages), name and title of the approving official, and date of approval.

As for the format of the written method on file, Dux suggests including a method number, date authorized, references to supporting literature, method scope and basic principles, apparatus and reagents required, safety precautions, detailed procedural steps, calculation formulas, information on precision and accuracy, reference sample QA check requirements, and any appropriate special comments. The AOAC recommends a somewhat similar standardized format to its members (*23*).

Accuracy and Precision

The terms related to error, accuracy, and precision may require better definition (*24*). *Errors* can be classified as systematic (or determinate) or random (indeterminate). *Systematic errors* have a variety of causes, such as defects in the method, poor analyst practices, instrument malfunction, and invalid standards. These errors are usually constant for a given method and can be accounted for or measured. *Random errors* are unavoidable and are reflected in the unpredictable fluctuations that occur in the use of a method. They can be described using statistical procedures.

Accuracy, of course, is the closeness of a result, or the arithmetic mean of a set of results, to the true, expected, or accepted value. *Precision* is the agreement or repeatability of a set of replicate results among themselves or the agreement among repeated observations made under the same conditions. Measures of precision are qualified or explained in terms of possible sources of variability: replicability, repeatability and reproducibility. The following chart reflects the requirements of the definition (*25*):

Source of Variability	Replicability	Repeatability	Reproducibility
Specimen (Subsample)	Same or Different	Same or Different	Most Likely Different
Sample	Same	Same	Same
Analyst	Same	(At least one of these must be different)	Different
Apparatus	Same		Different
Day	Same		Same or Different
Laboratory	Same	Same	Different

The actual performance of a method is the real test of its applicability to the situation at hand. For some methods, when the results are critical and the method is intermittently, but regularly, used by several analysts, it may be desirable to have the method studied for repeatability. This involves the examination of duplicate samples at staggered times. It may also be desirable to include in the study more than one analyst using the same or different instruments. The following chart can be used as an example of a schedule for the performance of the analyses in a test for repeatability (25):

Fri	Mon		Tue		Wed		Thu		Fri	
PM	AM	PM	AM	PM	AM	PM	AM	PM	AM	PM
O1	O2			O3	O4			O5	O6	
		D1		D2	D3			D4	D5	D6

O — Original sample analysis and test number
D — Duplicate analysis on corresponding original analysis

Critical Control Points

Some laboratories, as part of their quality assurance programs, establish critical control points and specifications for individual methods that are to be given particular attention in their analytical schemes. The following example, taken from the Quality Assurance Manual of the Food Safety and Inspection Service (FSIS) of the U.S. Department of Agriculture illustrates this practice (26):

In the analysis of PCBs, using the AOAC's "Organochlorine and Organophosphorus Pesticides Residues" method (27), several critical control points are recognized. Under **970.52B**, "Solvent Purity Test," the AOAC method states "Electron capture GC requires absence of substances causing detector response," using a test described. The FSIS then adds "No deflection >3 mm from base line from 2–60 minutes after injection of concentrated solvent." Under **970.52B(i)**, "Florisil," the AOAC method states in part "60/100 PR grade, activated at 675°C (1250°F)." The FSIS then adds "Activated at 1250°F ≥2 hours if not already done by the supplier followed by ≥5 hours at 130°C and storage in desiccators. ≥5 hours at 130°C must be repeated for each batch at least every 2 days."

The FSIS also suggests that a PCB screening procedure be used, such as gel permeation chromatography or a micro alumina column procedure. In these procedures the following standards must be met: (a) It must not miss any PCB residues present at or above the detection limit; (b) it must not report any false positives; (c) the recovery for PCB residues must be 60 to 100% with a CV<20%; (d) values found, corrected for recovery, must agree within 20% with the recovery corrected values obtained with the official method; and (e) there must be documented validation that the extraction procedure used removes compounds of interest from the sample matrix.

The preparation of such a quality control system, which involves identification of critical control points for each method, can be quite time-consuming if the laboratory uses a wide variety of methods. On the other hand, the control can be effective in reducing annoying errors in analyses, especially when the analyte being measured is present in micro amounts.

Analysts

It is reasonable to expect an analyst, after proper training and appropriate experience, to produce reliable results in sample analysis and to operate with a minimum of supervision. This type of satisfactory operation can be achieved and maintained when appropriate methods and equipment are available, workload is reasonable, and adequate training is provided to prepare the analysts for new situations.

It is essential that even competent, experienced analysts familiarize themselves with new procedures and with seldom-used complex procedures in order to consistently produce satisfactory results. This is especially true as the analyte concentrations decrease to levels near zero, and as matrix complexity increases. Examination of intralaboratory check samples and interlaboratory proficiency check samples can provide a measure of an analyst's ability to cope with the newer methods.

Analysts must be trained to understand the limitations of their analytical findings. Calibration of the instrument and its precision and bias, and interferences caused by other constituents in the sample, represent the most common influences on a method's performance and account for much of the uncertainty associated with the analytical results produced. The quality of the written description of a method influences how well it will be interpreted and can make a serious contribution to the overall error in the measurement process. How well the analyst understands the written procedure will depend not only on how well it is written, but on the analyst's ability to understand written technical procedures, as well as specific knowledge and experience as an analyst. In any event, the analyst is expected to use good judgment in the interpretation of the method, and then *follow the method as closely as possible.*

Analysts need to understand that their results can be distorted by interferences contributed either by the sample itself or by the procedure used to analyze the sample. These latter effects are referred to as nonmatrix effects, and include such things as the contributions from contaminated reagents or glassware. Instrument noise can also play a part in the production of misleading data. Both matrix and nonmatrix effects have an especially significant influence in analyses of trace quantities of analyte.

In some instances the sample material itself can cause problems that are manifested in different ways. The most common of these is an apparent increase in the analytical response as a result of an irregularity of the baseline from which the analyte response is measured. The sample matrix, of course, can affect the recovery of the analyte, either increasing it or decreasing it in any given situation, and the "cleanup" that precedes the determinative step, by varying in efficiency from run to run, will change the composition of the final matrix and therefore any influence the matrix may have on the determinative step.

The FSIS Quality Assurance Program mentioned earlier requires a certain level of performance from analysts: a readiness to perform, examination of intralaboratory check samples, and participation in interlaboratory check sample programs. In addition, "acceptability criteria" are established for complex analyses. If the analyst fails to maintain an acceptable level of competence, investigative and corrective actions are taken. Again using the PCB analysis as an example, FSIS requires of analysts recoveries of greater than 75%, that the coefficient of variation be less than 12%, that no false positives be found, that no PCB residues be missed if present above their lowest detectable limits, and all reported values, corrected for recovery, be within 1.7 standard deviations (90% confidence) of the average value, corrected for recovery, found by all laboratories that have analyzed the same samples. Somewhat similar acceptability criteria are established for intralaboratory check samples.

The Environment

The performance or effectiveness of people, instruments, equipment, and even reagents are affected by environmental conditions. For example, uneven room-temperature throughout the day in a laboratory will severely affect the retention times of room-temperature chromatographic procedures such as high pressure liquid chromatography. If analyses are being performed over a period of an hour or longer, the sample and standard chromatograms may not be comparable because of temperature fluctuations. Uneven room temperatures from spot to spot in a laboratory can also affect testing since, for example, temperature changes affect volume measurements of organic liquids. Humidity, as mentioned earlier, can affect samples that are opened in the laboratory—they can either gain or lose moisture rapidly depending upon the atmospheric conditions. Certain reagents are well known for taking up moisture, and fine powders that have considerable surface area exposed can adsorb moisture in a humid environment or lose moisture to a dry one. Certain chemicals are sensitive to the ultraviolet light contained in fluorescent lighting, and for those laboratories that still have windows, sunlight can be an important environmental "contaminant." Finally, the category of potential environmental problems also includes the utilities. Examples are spikes or irregularities in the electrical power, contaminants or temperature fluctuations in the water supply, and pump oil or other contaminants in the laboratory air supply.

Instruments

Since instruments consist of mechanical, optical, or electronic components, their performance tends to change as the these components begin to age or wear. It is difficult to detect slow changes in day-to-day or even week-to-week operation, and since most measurements are made by comparing the response of the sample with the response of a standard, evidence of changes in the absolute response of the instrument tend to be overlooked (28). If these subtle changes are not monitored, problems will eventually occur. As a general rule, instrumental errors of this kind are controlled through a consistent practice of periodic calibration.

The most common instrument standardization techniques are the *analytical calibration curve,* the method of standard additions, and the internal standard method. The analytical calibration curve is the most common and most frequently used method. In this method the signal response of the instrument is determined using various concentrations of analyte in a suitable solvent or matrix, and the relationship is graphed. The *standard additions* technique is used when it is not possible to suppress interferences from elements in the matrix. As explained earlier, the technique is more involved than the usual calibration curve, but it is sometimes the only alternative when matrix effects create problems. The method of *internal standards* uses the addition of a fixed quantity of a secondary standard which permits its measurement and comparison with the analyte but does not interfere with the response of the analyte. The method of standard additions or of internal standards generally yields more accurate results than the use of the simple calibration curve based upon external standards.

Calibration intervals, that is, a schedule that requires that specific performance tests be made on each instrument, must be assigned to all equipment as part of the laboratory's preventive maintenance program (see Chapter 4, Management of Equipment and Supplies).

Although preventive maintenance does provide greater assurance of the reliability of results, it does not always completely control equipment performance. In addition to regular preventive maintenance, periodic reconditioning of instruments is important, as is day-to-day care in the basics of instrument use. A few examples of common technical mistakes in instrument operations and use are cited below (29):

- Failure to check the level of an analytical balance, failure to allow objects to reach thermal or humidity equilibrium before weighing, or failure to dissipate static charges can cause problems in weighing.
- Some instruments require an adequate warmup time to reach stability, and others need constant monitoring of electronic drift; failure to observe these conditions can cause errors.
- There can be relative instrument errors associated with absorbance measurements, for example, when spectrophotometers are used outside of their optimum performance range, often limited to 0.2 to 0.7. The stability of some instruments tends to change with time; therefore samples and standards must be measured at relatively narrow intervals.
- Unstable baselines can lead to measurement errors. Improper judgment of how to interpret or extrapolate a baseline similarly can lead to inaccurate results.
- Failure of any of the components in a chromatograph to function at optimum, such as the oven temperature, gas or liquid flow system, detector, electrometer, or recorder, can distort overall instrument response.
- Failure to condition a chromatographic column or to conduct an evaluation of its quality characteristics may fail to reveal its poor performance or complete inadequacy to properly resolve and measure compounds.

Mistakes (Blunders)

Simple mistakes (or blunders) are easy to make during analyses, but of course can do major damage. They can be minimized by training and by close attention to the assignment by supervisory and operating personnel. The following list, taken from a variety of sources, represents mistakes frequently encountered in laboratory operations:

- Nonrepresentative sampling
- Incorrect identification of samples
- Failure to follow method directions
- Failure to report important observations or information
- Mistakes in reading instrument data or volumetric measurements
- Incorrect recording of data, such as transposition of digits, incorrect location of decimal points, or confusing or inverting numerator and denominator
- Improper reduction of analytical data or errors in reduction of information from charts
- Failure to purify reagents or to use those of suitable quality
- Incorrect standardization or failure to restandardize solutions (this is one of the most frequent causes of error)
- Incorrect reading of retention times or peak heights
- Use of glass cells in the UV region
- Improper dilutions or incorrect reporting of dilutions
- Failure to use matched cells
- Use of inadequately cleaned glassware
- Incorrect readings of injection volumes
- Errors in calculating results
- Incorrect interpretation of data
- Mistakes in transposing data between records or between the back and the front of a worksheet
- Incorrect keying-in of data into computers
- Careless handling of tapes or discs

Poor handwriting has always been a problem but has become a growing one. In the laboratory, inaccurate typing by analysts is a relatively new problem. Numerals carry so much information in so little space that it is especially important that they be carefully formed, or typed, and *never* overwritten. Changes should be made only by crossing out, initialling the cross-out if it represents original data, and rewriting, legibly, the correct numeral. Typed numerals must always be carefully double-checked by the analyst who typed them.

Control Charts

Laboratories that analyze large numbers of the same type of samples can find analytical control charting a very useful tool in ensuring the effectiveness of their quality control efforts. These charts display data in a form that compares the variability of test results with the average or expected variability of small groups of data. The results of analyses are plotted on the vertical axis of a chart, in units of the test results, against time on the horizontal axis (30). When the chart covers a sufficient period, it provides a ready means to determine trends or the lack of randomness (bias). The use of control charts, how they are prepared, and how they are interpreted are discussed further in Chapter 2, "Statistical Applications and Control Charts."

Corrective Actions

As part of the quality assurance program, it is desirable to develop a plan for taking prompt corrective action when errors, deficiencies, or out-of-control situations are observed. There are two types of corrective actions: on-the-spot or immediate action and long-term action. On-the-spot action is used to correct minor problems, such as the repair of a malfunctioning piece of equipment that requires replacement of a minor part, or the correction of poor analytical technique being used by an analyst. *On-the-spot actions* are taken frequently. They involve situations that can be handled promptly by the analyst, or dealt with by the supervisor, and brought to the attention of an individual analyst or to the entire staff if it appears that the problem may be widespread. *Long-term corrective action* is used when an out-of-control or highly unsatisfactory condition is discovered and its correction requires careful consideration and involves a number of individuals. A general improvement of data quality, for example, or the need to make major repairs on an instrument or major corrections to a laboratory procedure would require long-term corrective action.

In a small laboratory, or one in which few problems are encountered, the corrective action can be casual. When problems are numerous or severe, a more formal approach is necessary, such as a "closed-loop" corrective action system (31). The steps in such a system are to define the problem, assign responsibility for investigating the problem, investigate and determine the causes, develop corrective action to eliminate the problem, assign responsibility for implementing the corrective action, establish the effectiveness of the corrective action and implement the correction, and verify that the corrective action has eliminated the problem. This system involves the use of forms to document the various steps (see Appendix B, Figures 10 a,b,c, for suggested forms).

This system may be too involved for some laboratories and perhaps unnecessary for small laboratories, but regardless of the scheme used, corrective actions must be a continuing part of a laboratory's plan for quality assurance in sample analysis.

Recommendations

As part of the management and control of sampling, sample preparation, and sample analysis, the following actions are recommended:

1. Work with appropriate persons to develop sampling plans for the various types of products delivered to the laboratory for analysis.
2. Establish subsampling procedures for various products, giving consideration to the use of composites or individual unit examinations, based on the variability to be expected among sample units and the resources available for their analysis.
3. Prepare guidelines for sample preparation for analysis that will minimize composition change.
4. Select methods of analysis for which performance data are available that suits the purposes of the laboratory, giving consideration to the suggestions made in this chapter.
5. Subject all nonofficial or modified methods to validation testing to ensure their suitability with respect to accuracy and precision for the type of sample under consideration.
6. Conduct regular control checks to determine the continuing satisfactory performance of the analytical control program.
7. As deemed necessary, subject certain methods to critical analysis and establish critical control points to minimize errors.
8. Maintain constant supervisory vigilance to minimize inadvertent processing and operational errors.
9. Supervise basic instrument use, in addition to preventive maintenance, to avoid or minimize errors arising from improper analytical technique.
10. Provide familiarization time for analysts when new or complex procedures or instruments are used in sample examination.
11. Set acceptable performance-level criteria to maintain quality levels of competence for analysts.
12. Use control charts to maintain statistical control for repetitive-type analyses.
13. Take on-the-spot or long-term corrective actions, as necessary, to correct or respond to instrument failure, poor analyst technique, or out-of-control or unsatisfactory performance.

References

(1) Whitaker, T.B., et al. (1974) *J. Am. Oil Chem. Soc.* **51**, 214
(2) "Sampling Procedures and Tables for Inspection by Attributes," MIL–STD–105D (1963) U.S. Naval Supply Depot, Philadelphia, PA
(3) "Sampling Procedures and Tables for Inspection by Variables for Percent Defective," MIL–STD–414 (1957) U.S. Naval Supply Depot, Philadelphia, PA
(4) Horwitz, W., et al. (1977) in *Quality Assurance Practices for Health Laboratories,* S.L. Inhorn (Ed.), American Public Health Association, Washington, DC, p. 559
(5) The following may be found useful:
   - Horwitz, W. (1988) *J. Assoc. Off. Anal. Chem.* **71**, 241–245
   - Springer, J.A., & McClure, F.D. (1988) *J. Assoc. Off. Anal. Chem.* **71**, 246–250
   - Park, D.L., & Pohland, A.E. (1989) *J. Assoc. Off. Anal. Chem.* **72**, 399–404
   - Garfield, F.M. (1989) *J. Assoc. Off. Anal. Chem.* **72**, 405–411
(6) "Agricultural Food Products — Layout for a Standard Method of Sampling from a Lot," ISO/TC34, ISO/DIS7002.2 (1988) *ISO Catalog,* International Organization for Standardization, Geneva, Switzerland
(7) Youden, W.J., & Steiner, E.H. (1975) *Statistical Manual of the AOAC,* Association of Official Analytical Chemists, Arlington, VA, p. 41
(8) "Improving Analytical Chemical Data Used for Public Purposes" (1982) *Chem. Eng. News* **10** (23), 47
(9) Benedetti-Pichler, A.A. (1956) *Essentials of Quantitative Analysis,* The Ronald Press Co., New York, NY, p. 319
(10) Crosby, N.T. (1986) "Quality Assurance in Sampling," presented at an AOAC–LGC short course on "Quality Assurance for Analytical Laboratories," Sept. 1–3, 1986, London, England, Association of Official Analytical Chemists, Arlington, VA
(11) Campbell, A.D., et al. (1986) *Pure and Applied Chem.* **58**, 205
(12) *Official Methods of Analysis*, 15th Ed. (1990) Association of Official Analytical Chemists, Arlington, VA, p. 274
(13) Taylor, J.K. (1987) *Quality Assurance of Chemical Measurements,* Lewis Publishers, Chelsea, MI, p. 78
(14) *Handbook for AOAC Members*, 5th Ed. (1982) Association of Official Analytical Chemists, Arlington, VA
(15) Horwitz, W. (1982) *Anal. Chem.* **54**, 67A–76A
(16) Cobb, W.Y. (1986) "Quality Assurance in Quantitative Analytical Determinations," presented at an AOAC Quality Assurance Short Course, Arlington, VA 1986, Association of Official Analytical Chemists, Arlington, VA
(17) Crable, J.V., & Smith, R.G. (1975) *J. Am. Ind. Hyg. Assoc.* Feb., 149

(18) Egan, H. (1974) *Report of the Government Chemist — 1973,* Her Majesty's Stationary Office, London, England

(19) McCully, K.A., & Lee, J.G. (1980) in *Optimizing Chemical Laboratory Performance Through the Application of Quality Assurance Principles,* F.M. Garfield et al. (Eds), Association of Official Analytical Chemists, Arlington, VA, p. 73

(20) Dux, J.P. (1986) *Handbook of Quality Assurance for the Analytical Chemistry Laboratory,* Van Nostrand Reinhold Co., New York, NY, p. 36

(21) Youden, W.J., & Steiner, E.H. (1975) *Statistical Manual of the AOAC,* Association of Official Analytical Chemists, Arlington, VA, p. 33

(22) Dux, J.P. (1986) *Handbook of Quality Assurance for the Analytical Chemistry Laboratory,* Van Nostrand Reinhold Co., New York, NY, p. 46

(23) *Handbook for AOAC Members,* 5th Ed. (1982) Association of Official Analytical Chemists, Arlington, VA, p. 37

(24) The following may be found useful:
- Fitz, J.S., & Schenk, G.H. (1966) *Quantitative Analytical Chemistry,* Allyn and Bacon, Boston, MA
- Currie, L.A. (1978) *Treatise on Analytical Chemistry,* I.M. Kolthoff & P.J. Elving (Eds), Wiley Interscience, New York, NY
- Westgaad, J.O., & Hunt, M.R. (1973) *Clin. Chem.* **19**, 49
- "Standard Recommended Practice for Use of the Terms Precision and Accuracy as Applied to Measurement of a Property of a Material," ASTM E177 (1971) American Society for Testing and Materials, Philadelphia, PA

(25) *Quality Assurance Handbook for Air Pollution Measurements,* Vol. 1 (1984) U.S. Environmental Protection Agency, Research Triangle Park, NC, Sec. 1.4.9

(26) *Chemistry Quality Assurance Handbook,* Vol. 2 (1982) U.S. Department of Agriculture, Food Safety and Inspection Service, Washington, DC, Sec. 2.5.10

(27) *Official Methods of Analysis,* 15th Ed. (1990) Association of Official Analytical Chemists, Arlington, VA, p. 274

(28) *Instrumentation Quality Assurance Manual* (1979) U.S. Consumer Product Safety Commission, Washington, DC, p. 9

(29) Sherma, J. (1976) *Manual of Analytical Quality Control for Pesticides in Human and Environmental Media,* U.S. Environmental Protection Agency, Research Triangle Park, NC

(30) *Quality Assurance Handbook for Air Pollution Measurements,* Vol. 1 (1984) U.S. Environmental Protection Agency, Research Triangle Park, NC, Appendix H

(31) *Quality Assurance Handbook for Air Pollution Measurements,* Vol. 1 (1984) U.S. Environmental Protection Agency, Research Triangle Park, NC, Sec. 1.4.13

Chapter 7

PROFICIENCY AND CHECK SAMPLES

"How do you know your results are right?" This interesting question was addressed by Horwitz, who pointed out that analysts used to have "full control and responsibility over the production of the data at every step. He or she prepared reagents, calibrated weights and volumetric glassware, and standardized the output of instruments" (1). Now, with prepared reagents, glassware often washed, stored, and distributed by a special unit, calibration of instruments assigned to the manufacturers, proper functioning of instruments assumed to be built in by instrument designers, and so forth, the production of data is no longer under the total supervision of the analyst. Production of the data is a "maze-type" function in which machines "measure the samples, execute the manipulations, determine the response, perform the calibrations, and present the final answer in whatever form or units are desired."

The obvious answer to Horwitz's question is reference to a reliable quality assurance program. Horwitz reaches the same conclusion, and in his closing paragraph states, "The most difficult part of the procedure of producing reliable analytical values will be obtaining recognition by analysts of the necessity for quality control as an inherent accompaniment of analytical work."

Results validation, an important control step in the quality assurance program, provides a check on the laboratory's analytical results. Failure to consider this measure can leave a crucial void in the program. This check is in addition to the everyday monitoring of activities by supervisors. Results validation, or proficiency testing as it is termed by some, is a systematic testing program in which samples of known composition are analyzed to evaluate a laboratory's ability to perform at an acceptable level of competence. This type of testing includes both intralaboratory and interlaboratory sample examinations.

Intralaboratory Testing

The purpose of an intralaboratory testing program is to provide a periodic assessment of the performance of individual analysts and of the overall laboratory. The testing program also provides the means to establish a record of instrument performance, demonstrate the reliability of analytical methods, detect training needs, and highlight any needs for upgrading laboratory quality performance. Intralaboratory performance testing includes the examination of check or reference, "blind," and "double-blind" samples, and extensive use of primary and secondary standards.

There are three questions to be resolved in an intralaboratory program: what kinds of samples should be used, how should these samples be prepared and introduced into the daily routine, and how often should these checks samples be run. The kinds of samples to use will depend on the kinds of samples examined by the laboratory. When a variety of commodities are analyzed, and several different determinations are made on each, the decision is not an easy one. If a given type of sample is or will be examined on a regular basis, a reserve portion of a sample can be retained as a check sample once its composition is established and the analytical process is demonstrated to be in statistical control. One or more fabricated test samples can also be prepared and used.

There is some disagreement among quality assurance people as to the use and value of blind or double-blind samples. A blind sample is one whose composition is known to laboratory management, but unknown to the analyst. A double-blind sample is one whose composition is known to the laboratory management, but neither its composition nor its identification as a check sample is known to the analyst. In this latter situation the sample is assigned to an analyst who believes it to be a routine sample. One expert believes that the advantage of the double-blind sample is that it can be used to evaluate the true quality of the system (2). It has been demonstrated repeatedly that the variance on a double-blind sample is greater than that on a sample that is identified as a check sample. Analysts tend to be more cautious in their approach to known test samples and will often repeat analyses until they are confident of the results reported. Another expert thinks that the value of blind and double-blind samples may be overplayed and that excessive use of them can be counterproductive, especially if it engenders an air of distrust among analysts (3). He also thinks that, when good control is achieved over a period of time, the frequency of such samples can be reduced.

There seems to be little question that blind sample testing has advantages, especially when analysts understand the reasons for the program. On the other hand, the time and effort required may be more than the laboratory can support. For commercial testing laboratories, the procedure may be less difficult and more desirable or even necessary to establish the credibility of the laboratory's performance. How often to check the performance of analysts is a judgmental decision. The frequency will depend on the volume of work, the experience of the analysts, costs involved, and the risk that management is willing to take that the analytical system is in control.

As a general rule, regulatory agencies require, before legal action is considered, a check analysis by a second analyst of any sample that is found to be outside required limits. Usually a second method, based on a different principle, is used in the check determination. Samples that do not indicate a probable need for regulatory action normally are not subjected to check. Some laboratories may not be large enough to have a second analyst available with the necessary expertise; in this situation the laboratory may use the original analyst to reexamine the sample, using a second method, if possible.

For pesticide residue samples, Watts (4) suggests that when laboratories make less than one routine analysis per week of a given product they should analyze a corresponding test sample with each routine sample, and not less than one test analysis per month if no routine sample is encountered. If more than one routine sample per week is examined, then at least 10% as many test samples as routine samples should be run, with a minimum of one per week. He thinks that no special care should be given to test samples; they should be carried through the analysis concurrently with other samples.

Horwitz et al. (5) state that between-analyst, within-laboratory precision is useful in the assessment of analyst performance. It shows how closely two analysts in the same laboratory using the same method can check each other. They should certainly do better than two analysts in different laboratories. Horwitz and co-workers observe that the "difference in values between analysts in the same laboratory on the same sample (a function of between-analyst precision) can be conveniently monitored by control-chart techniques. The analysis of butter for fat or of tuna fish for mercury are candidates for this type of quality control."

Interlaboratory Test Programs

Interlaboratory test programs are identified as proficiency testing, interlaboratory surveys, check samples, or round-robin programs. In these programs, samples are distributed for analysis to laboratories by organizations that are external to the laboratories. These are usually voluntary participation programs used as part of a system of laboratory accreditation or certification, or to estimate the accuracy and precision of results between laboratories, or to upgrade the overall quality of laboratory performance. The purpose of the interlaboratory testing program is normally stated in the literature of the sponsoring organization.

Another form of proficiency testing is used by some government agencies as part of their programs for accrediting laboratories for conducting certain types of tests (6). Comparison or split samples are analyzed by both the laboratory seeking accreditation and the government laboratory, with the result from the government laboratory being the standard. The laboratory being accredited may be required to maintain quality control charts, and each analyst may be required to maintain individual charts. These findings are reviewed by the government's quality control officer to ascertain whether or not the analytical work meets established performance standards. This proficiency testing is only a part of the requirements for accreditation.

Collaborative testing is a special form of testing to evaluate the performance of a method under actual working conditions in several laboratories through the analysis of carefully prepared homogeneous samples (7). Basically, it is not intended as an assessment of analyst performance, but whether the method being studied is in fact a reliable one. Nevertheless, participation in collaborative studies can provide some information that can be used for analyst results assessment in areas of interest to the laboratory.

There are a number of advantages to participation in an interlaboratory testing program (8): It provides a means to compare laboratory performance with that of other laboratories; it can encourage self-appraisal and a minimizing of laboratory errors; it provides external evidence of the quality of the laboratory's analytical performance and individual analyst proficiency; and it can be used to reduce the frequency of intralaboratory testing when consistently favorable results are achieved on interlaboratory test samples. Furthermore, in common with other parts of the quality assurance effort, interlaboratory testing can assist in identifying needs for training and changes in laboratory procedures.

Proficiency Test Program Format

Voluntary interlaboratory proficiency testing programs generally have a coordinating organization or laboratory that provides samples to be examined by participating laboratories. This type of program has the following features (9):

- The coordinating laboratory prepares a homogeneous sample and distributes portions to participating laboratories for analysis.
- The sample may or may not be accompanied by a standard method. If a standard method is not used, the analyst is usually requested to describe the method used.
- The analyst is required to examine the sample within a specified time and to submit the results to the sponsoring organization, through the laboratory director.
- The sponsoring organization collates the data, performs a statistical analysis, and sends a report to the participating laboratories. A relative performance or ranking scheme may also be prepared, establishing a numerical ranking for each laboratory, or a qualitative ranking.
- In cases of poor performance, the laboratories are contacted, and suggestions may be made for improving performance.

Laboratories participating in the test are usually identified by code so that their performance remains anonymous to other participants. The analyst is not named, and in some laboratories the analyst may not know that he or she has examined a proficiency test sample until some later time.

There are several major components of a proficiency test program that must be controlled: the sample, the analytical method, analytical instruments, standards, and critical reagents (10). Unless these elements are carefully controlled, the results of a study may be of questionable significance. The purpose of the program is to show laboratory and analyst variability, so other variables must be carefully excluded. If a particular result is out of line, attention is given to correcting those factors that were not controlled, namely, the analyst and the laboratory environment.

The sample is probably the most important element in the process because it is common to all participants. Samples must be uniform with respect to the concentration, identity, and distribution of the ingredient that it is to be analyzed for, and the samples must be carefully packaged and shipped in order to avoid changes before the analysis is made. To ensure this, the issuing laboratory will normally prepare and package the sample, then analyze several portions themselves. It is a small cost for avoiding the tremendous waste of time and money that would occur if a faulty sample were issued. The portions delivered to each laboratory generally have a variation less, "by an order of magnitude, than the variation stemming from the analytical differences between laboratories" (*11*). Essentially, three procedures are used to establish the "correct" or "true" value for the test sample: (a) authentication by the laboratory that produces the test sample, (b) authentication by reference or referee laboratories, and (c) authentication by participants.

Authentication by the producer of a test sample may seem simple, but it is not. Preparing a homogeneous sample is most difficult and requires great care. A vegetable oil sample containing pesticides, for example, may be quite uniform, but a solid that is spiked with pesticides presents a difficult problem.

Use of a single laboratory for authentication of a test sample is usually unsatisfactory. If the laboratory is of the caliber of the National Institute for Standards and Technology, its authentication may be sufficient, but few laboratories fall into this class, and the number or variety of samples that a single laboratory can prepare is limited. Another approach is the use of several referee laboratories. These laboratories are recognized as the elite establishments in the field. Their results then become the basis for the "correct" answer. Authentication by participants is widely used, in which the average finding of a number of laboratories is the assumed correct value, but it may not be a reliable strategy unless its limitations are recognized.

Special instructions may have to accompany the check sample with respect to the quality or grade or even the manufacturer of materials required for the analysis. In some cases, certain reagents may have to be distributed to participants by the organization coordinating the proficiency test. Special care must be given, for example, to chromatographic columns, chromatographic media, special solvents, and reference standards. Obviously, the method used will have a great influence on the results. If the method is not specified and various methods are used, the test results will probably be less accurate and precise than they would be if a "standard" method were used, and without this control it will be difficult to conclude where the problem lies when an unsatisfactory result is obtained. Even if a specified method is used, the results may vary significantly if the method has not been successfully subjected to a collaborative study. Ideally, the method specified will be one that is fully validated and familiar to all participants.

The purpose of proficiency test samples is to determine whether the laboratory is capable of correctly performing analyses that it is called upon to perform. Although some proficiency test samples are preceded with one or more preliminary rounds to familiarize analysts with the method, and to work out any problems, in principle this defeats the purpose of the program, unless it is the rigid practice of that laboratory to do training recovery runs before analyzing *any* sample of the type represented by the proficiency sample.

Instrument control is essential and can usually be achieved by routine calibration and the use of reference standards. Preventive maintenance, with regular instrument checks, and system suitability requirements built into the specified analytical method can eliminate a variable that is often difficult to control.

The analyst is an important element in the proficiency test process. If the laboratory is interested in making a "best" impression, an experienced analyst will probably be selected to examine the test sample. This procedure results in a biased situation because the sample is not treated in the way that routine samples are. Further, if the sample is recognized as a proficiency test sample, the analyst will most likely give it extra care. In a study conducted by one Federal agency, analysts located in different laboratories, after an extended period of practice, were able to analyze a standard set of samples with an acceptable standard deviation of their pooled measurements. When these same samples were examined as routine unknowns, however, the number of analysts that obtained acceptable results fell by 50%. This clearly indicates that the analysts took extra care when they knew they were participating in a proficiency check program — in short, they performed better when their work was under scrutiny. Thus the assignment of the sample is an important administrative problem. The director of the laboratory must decide whether "best" results are desired or knowledge of how well the laboratory performs its routine operations. In some cases, this decision will be governed by the purpose of the particular program.

Environment may have a significant effect on a test sample just as it can on any sample. It is incumbent on the analyst to be aware of contamination possibilities for the *particular sample being analyzed* and to take necessary steps to maintain the integrity of the sample. Examination of samples for trace amounts of pesticides may be compromised if macro amounts of the same pesticides are examined in the same laboratory. When analyzing for glass fragments, avoid glassware; when analyzing for particulates, avoid tools and apparatus that could shed particulates. Failure to clean glassware and work areas adequately can easily lead to erroneous results.

All of these precautions are really part of laboratory quality assurance and some of the more significant problems of laboratory management. Recognition of these potential problems is a necessary step toward success in a test program as well as in the production of reliable data on a routine basis.

According to Watts (*12*), during the early years of the interlaboratory quality control program coordinated by the U.S. Environmental Protection Agency that was involved with determining pesticides and related chemicals in human tissues, fluids, and environmental media, the results were extremely poor. Results improved when uniform methods and standardization among all laboratories were achieved. For example, the relative standard deviation (RSD) and coefficient of variation (CV) on test samples of fat for pesticide residues in 1967, involving 15 laboratories, was 50%. In 1980, with 12 laboratories, the RSD was 20% on the same type of check sample, a significant improvement. Similar improvement occurred in other types of samples, and the improvement was maintained over the years the program was in existence.

Test Programs in the United States

Proficiency testing, as an instrument of evaluation or regulation of chemical laboratories and for many industrial health laboratories, is included in the regulations of several states. It has been made a requirement for voluntary accreditation by programs such as the Joint Commission on the Accreditation of Healthcare Organizations, the College of American Pathologists, and the American Industrial Hygiene Association. The concept of proficiency testing was incorporated into the U.S. Medicare Act of 1965 and was further refined in 1974 in the revised conditions for coverage of independent laboratories. It is also covered in the U.S. Clinical Laboratories Improvement Act of 1967 (*13*).

Several Federal agencies, such as the Food and Drug Administration, the Drug Enforcement Administration, and the Department of Agriculture, each of which has multiple laboratories, conduct their own check sample programs.

A number of associations and agencies furnish check samples and operate proficiency test programs. Most programs offer samples on a regular basis, some more frequently than others. Commodities are offered in the areas of interest of the association, generally to member organizations. A fee is usually charged to support the program.

The Science and Technology Committee of the Association of Food and Drug Officials in June 1988 published an extensive list of "Proficiency and Check Sample Programs." The list identifies many active programs and is divided into two parts:

- Part 1 is a list of sponsoring organizations; and
- Part 2 is an alphabetical list of programs.

The list is reproduced in Appendix E. It is certain that this list will change from time to time with some sponsoring organizations added and others removed, but it can serve as a starting point for laboratories interested in participating in various programs.

Recommendations

As part of the laboratory's quality assurance program, it is suggested that the laboratory consider the following:

1. Establish an intralaboratory test program and require participation by all analysts.
2. Use the results of the intralaboratory test program to improve quality control, initiate training, and revise analytical procedures as necessary.
3. Participate in interlaboratory test programs in areas of the laboratory's interest.
4. Study the results of proficiency tests and move quickly to determine the causes for results that are not satisfactory.
5. Document results obtained in tests and make them part of the quality assurance program.

References

(1) Horwitz, W. (1981) *The Pesticide Chemist and Modern Technology,* ACS Symposium Series No. 160, American Chemical Society, Washington, DC, p. 411

(2) Dux, J.P. (1986) *Handbook of Quality Assurance for the Analytical Chemistry Laboratory,* Van Nostrand Reinhold Co., New York, NY, p. 28

(3) Taylor, J.K. (1987) *Quality Assurance of Chemical Measurements,* Lewis Publishers, Inc., Chelsea, MI, p. 146

(4) Watts, R.R. (1980) in *Optimizing Chemical Laboratory Performance Through the Application of Quality Assurance Principles,* F.M. Garfield et al. (Eds), Association of Official Analytical Chemists, Arlington, VA, p. 87

(5) Horwitz, W., et al. (1977) in *Quality Assurance Practices for Health Laboratories,* S.L. Inhorn (Ed.), American Public Health Association, Washington, DC, p. 545

(6) Barth, H.J. (1980) *Testing Laboratory Performance: Evaluation and Accreditation,* NBS Publication 591, G.A. Berman (Ed.), National Institute of Standards and Technology, Gaithersburg, MD, p. 169

(7) *Handbook for AOAC Members,* 5th Ed. (1982) Association of Official Analytical Chemists, Arlington, VA, p. 35

(8) Sherma, J. (1976) *Manual of Analytical Quality Control for Pesticides in Human and Environmental Media,* U.S. Environmental Protection Agency, Research Triangle Park, NC, Sec. 3

(9) Horwitz, W., et al. (1977) in *Quality Assurance Practices for Health Laboratories,* S.L. Inhorn (Ed.), American Public Health Association, Washington, DC, p. 616

(10) Bicking, C.A. (1980) in *Testing Laboratory Performance: Evaluation and Accreditation,* NBS Publication 591, G.A. Berman (Ed.), National Institute of Standards and Technology, Gaithersburg, MD, p. 31

(11) Forney, N.E., et al. (1977) in *Quality Assurance Practices for Health Laboratories,* S.L. Inhorn (Ed.), American Public Health Association, Washington, DC, p. 127

(12) Watts, R.R., (1980) in *Optimizing Chemical Laboratory Performance Through the Application of Quality Assurance Principles,* F.M. Garfield et al. (Eds), Association of Official Analytical Chemists, Arlington, VA, p. 96

(13) Forney, N.E., et al. (1977) in *Quality Assurance Practices for Health Laboratories,* S.L. Inhorn (Ed.), American Public Health Association, Washington, DC, p. 129

Chapter 8

AUDIT PROCEDURES

In the first chapter, quality assurance was defined as planned activities that are designed to ensure that the quality control activities are being properly implemented. The quality assurance program is a monitoring and auditing program that is designed to examine overall laboratory performance, evaluate it in detail, and disclose the cause of any less than satisfactory conditions. A successful quality assurance audit will lead directly to procedures that need changing, equipment that requires upgrading, analysts who are in need of training or better supervision, and so forth.

Performance Audits

The *performance audit* is a review, usually conducted by a supervisor, to evaluate the analytical activities of an analyst, as well as the data produced by that analyst. There are a number of separate reviews that can be considered part of the overall performance audit: worksheet or notebook review, oral worksheet or notebook review, onsite analyst work review, check sample analysis, and a review of intra- and interlaboratory sample analyses. Performance audits are basically a check on the performance of analysts and are sometimes categorized as a quantitative appraisal of quality. One or more of these reviews can be conducted during the same audit period.

The *worksheet or notebook review* is a review of sample analytical reports, recorded on worksheets or in notebooks, for completeness and accuracy. This entails the examination of a predetermined number of analytical reports for each analyst over a given period, for example, a month or a quarter. The review covers all critical items that the laboratory standard operating procedure requires analysts to report. It also covers the correctness of calculations; the proper choice of the method and a full description or citation; whether method modifications are reported fully; legibility and orderliness of the data recorded and conclusions drawn; proper reporting of equipment and standards used; proper indication of weights and tares; clear indication of who did what if more than one analyst was involved; proper identification of charts and other documents; and timeliness of the work. This is a retrospective review of work that has left the laboratory, and is in addition to the everyday control reviews maintained by supervisors on a sample-by-sample basis.

The oral *worksheet or notebook review* is most common to organizations whose analytical data are sometimes used as court evidence. The review is similar to the one cited above, but it is a joint effort of the auditor and the analyst. The analyst is required to refer to data that is are least six months old, and from them reconstruct for the auditor the procedures that were followed and why those procedures were selected, and to explain the results reported. This has been found to be an excellent training device to impress upon the analyst the importance of thinking through each step and understanding why each choice was made. The oral review also emphasizes to the analyst the importance of making the analytical records clear and complete.

The purpose of the *onsite analyst work review* is to observe the analyst conducting typical analytical operations. Not every possible error is going to be evident from what the analyst places on a worksheet or in a notebook. This assessment provides the auditor an opportunity to observe the employee's analytical technique and to evaluate whether the analyst is following approved laboratory procedures. Care must be exercised by the auditor not to interfere with the analyst's work, or by actions, words, or facial expressions to communicate that the analyst either is or is not proceeding properly. The review period is generally one day.

The *check sample analysis review* is an actual repeat of sample analysis by the auditor or a second analyst. Unlike the check analysis referred to previously, this review can be performed on a sample that is either in limits or out of limits, and it is usually performed using the same method used by the original analyst, but, as with the check analysis, the purpose is to find out whether the first result was valid. This serves to check both technique and instrument stability.

Intra- and interlaboratory sample analyses review is a study of the results obtained by analysts in both intra- and interlaboratory proficiency test programs. It provides the means to assess individual analyst performance by comparing it with that of other persons in the same laboratory and other related laboratories. Its primary limitation is that ordinarily the number of check samples is not large.

System Audits

The system audit is an onsite inspection or assessment of the laboratory's control system. In comparison with the performance audit, it is sometimes referred to as a qualitative appraisal of quality. The system audit represents the traditional concept of quality assurance evaluation for laboratory accreditation or to provide evidence of a laboratory's competence. The audit requires the use of a quality assurance audit unit that applies carefully developed audit methods, evaluates their findings, and prepares a final report. Management must then use these findings to improve laboratory operations. The system audit can encompass the performance audits and will likely check on the work of the individuals, usually first-line supervisors, who conduct the performance audits.

The Quality Assurance Audit Unit

Many laboratories, especially larger multidiscipline laboratories, or multilaboratory organizations, maintain quality assurance audit units (1). The personnel of these units are responsible to management for monitoring the laboratory quality assurance program. It is desirable for the unit to be entirely separate from and independent of the personnel engaged in the direction and conduct of the laboratory being audited. One of the chief objectives of the unit is to highlight potential system problems that might be overlooked or missed by the daily reviews of the first-line supervisors or by the performance auditors. Another important objective, of course, is to improve laboratory procedures to achieve more efficient and accurate operations; in other words, ensuring that the laboratory maintains or, if possible, improves its capability for doing high-quality work.

It may not be possible in a small- to medium-sized laboratory to justify assigning full-time personnel to an audit unit. Under these circumstances, the persons selected to conduct a system audit should not be assigned to review their own work or work that is subject to the supervision of any of them. Regardless of the approach, the auditors must have enough scientific knowledge as well as audit training and experience to understand the work being audited and the requirements of the quality assurance program. With proper planning and experience, it is possible for two auditors to conduct a fairly comprehensive systems audit of a medium-sized laboratory (up to 20 analysts) in two to three days. It must be understood that this quality assurance unit is not responsible for *implementing* the scientific aspects of the quality assurance program, but for determining whether or not the program is an effective one, and the program is being followed.

Selection of Auditors

A successful audit program will be more likely to succeed if qualified persons are selected and trained as auditors. These people can exert a significant influence, positive or negative, on the quality assurance program and on scientific personnel. Although they often must be selected from a small group of available employees, their selection should be done as carefully as possible, taking into account such qualifications as education and technical competence, tact, self-confidence, thoroughness, objectivity, impartiality, ability to meet and deal with peers, and capability to communicate orally and in writing. A familiarity with and commitment to the quality assurance program is also highly desirable.

Auditor Training

Ideally, auditors should receive some training in what is expected of them. Normally, this is on-the-job training, beginning with an in-depth understanding of the laboratory quality assurance program, the organization of the laboratory, the laboratory mission and its typical workload, and general laboratory procedures. If the laboratory includes several scientific disciplines, it may be advisable for the audit team to include people experienced in each of the disciplines.

Auditors who may be laboratory supervisors themselves must divorce themselves from concern with their own quality assurance problems. They were assigned as auditors to do an important job, and must not behave in an apologetic manner for not being perfect in their own laboratories.

The training should conclude with a discussion between auditors and top management for whom the auditing program is being conducted. At this meeting the objectives and the procedural conduct of the audit will be reviewed, with both the auditors and management making contributions to what laboratory records and access will be needed, and what kind of close-out or briefing and final report will be expected.

Audit Planning

It is preferable that the audit unit operate from a comprehensive audit plan that is available in advance to all organizational units to be inspected. Just as any important laboratory procedure, the audit procedure should be in writing and widely understood. This open approach helps to ensure the objectivity of findings, provide credibility to the audit team, and serve to relieve tension of those being audited since the audit is cast as an important, but routine procedure, not singling out of particular analysts for evaluation.

In rare instances it may be desirable to perform unannounced audits. These permit auditors to determine how a laboratory operates from day to day when it is not prepared for the auditor's visit. There may even be some value in conducting an audit when the laboratory director is away. This situation provides information on how delegated management performs. These unannounced visits should not be frequent, however, and if they are to be used this should be spelled out in the written total audit plan.

The depth of the audit can vary considerably depending upon the size of the laboratory, the frequency of the audits, or the existence of any specific event having triggered the audit. The following types of information, however, are important for the auditors to have before the visit:

- Organizational structure
- Staff size, including number of supervisors, professionals, nonprofessionals, and clerical support
- Names, titles, educational background, and work experience of professionals
- Laboratory floor plan
- Workload, including types of commodities and numbers of samples of each commodity examined per year or in a given period
- Participation in proficiency or check sample programs
- Principal instruments and preventive maintenance programs for them
- Laboratory's quality assurance manual and all quality assurance program reports generated since the last audit

- Problem areas identified by their quality assurance program
- The previous two audit reports

The actual areas to be covered during the inspection should represent a cross-section of work being done by the laboratory. Each auditor should be assigned specific areas to audit and evaluate for the report, although all adverse findings should be discussed among the auditors. This will enable the team to understand how their responsibilities will be integrated for overall assessment of laboratory quality.

Audit Visit

The site visit normally begins with a meeting between the audit team, the laboratory director, and other responsible persons the director may invite. The following items are generally discussed: the purpose of the visit, the agenda for the evaluation, items and operations to be covered, items and records the audit team will require, and the criteria and guidelines to be used. The director is encouraged to comment on areas of concern that need attention. The team may wish to suggest a short staff meeting to explain the purpose of the visit and how long it will take. If the audit team is unfamiliar to the laboratory staff, it may help at such a meeting to introduce themselves and indicate their work and audit experience. The purpose of this approach is to relieve, insofar as possible, any fear associated with the audit and to create a more receptive attitude to a procedure intended to improve laboratory operations.

A checklist is useful to ensure complete coverage of important aspects of the audit. The checklist, when discussed with the laboratory director, will leave no open questions concerning the purpose or approach of the inspection. It should cover, at least, the following items:

- The quality assurance manual and the familiarity of laboratory personnel with it and with its purpose
- Methods identification, control, validation, and so forth
- Control of reagents, standard solutions, and their procurement and handling
- Sample control and chain-of-custody procedures
- Procedures for preventive maintenance for equipment and their conduct and control
- Analysis of samples and analytical reporting procedures
- Personnel manuals, personnel training, and personnel evaluation
- Ambient conditions, laboratory orderliness and cleanliness, safety practices, and the handling of toxic substances
- Intra- and interlaboratory proficiency studies
- Records preparation, storage, and disposition
- Deficiency handling and correction

It may be desirable, during the audit, to have the team, or each member, accompanied by a laboratory staff person. This helps to relieve tension, affords laboratory personnel the opportunity to observe at first hand the laboratory's operations from a quality control perspective, and to discuss administrative or operational problems encountered. All deviations from prescribed practices are noted for later discussion with the laboratory director.

After completion of the audit, the audit team, assigned staff persons, and the laboratory director meet for a briefing session. Objective observations by the audit team members are discussed, but subjective impressions are brought out as well, always with particular attention being given to problem areas. Oral recommendations to improve operations and the quality assurance program are offered by the team. This permits the laboratory group to offer its views and to comment on the audit team's recommendations. This exchange can serve to avoid "surprises" appearing in the final written report and unnecessary rebuttals based upon early misunderstandings by either the audit team or laboratory management. Credit is given by the audit team for high-quality performance found and for new ideas implemented by the laboratory that can be of value to other laboratories in the organization.

Evaluation and Reporting

Evaluation of the audit and preparation of a written report are as important as the actual onsite visit. The presentation of the list of findings, without assessment of their importance, is of little value. The report must indicate the degree of concern attached to any less-than-satisfactory conditions. Important defects require first attention and must be highlighted. Minor deviations also require attention, but can be corrected in appropriate time. Problems that are identified in the report must be documented with facts and observations, and subjective views held to a minimum.

The audit report must be prepared promptly and submitted to the audited facility as soon as possible, certainly within a month. Typically, at the top the report, the more important findings of the audit will be summarized. In the body of the report, proposed changes in laboratory procedures, including the quality assurance program, should be accompanied by solid reasons for such changes. Clear and straightforward language that is not unnecessarily critical will communicate best and limit the possibilities for misunderstanding or hair splitting. There should be no implications that individual employees are "violators," and the report, in so far as possible, should be constructive.

Before the report is put in its final form for submission to management, it is desirable that a draft be forwarded to the audited facility for review and a check of its factual accuracy. Frequently there are differences and disagreements over the findings that can be clarified to everyone's satisfaction at this point. Management may wish to have the laboratory submit its views before corrective actions are taken. The audit report and responsive comments are generally filed for future reference to see that corrections have been made.

FDA Approach to Quality Assurance

FDA has one of the largest field laboratory systems in the Federal government. These laboratories examine a large variety of commodities and operate in a number of disciplines. This agency has operated under a quality assurance program for many years and has taken a leading role in developing good laboratory practices. How FDA designed and implemented its program, and what it does with audit findings may be helpful to other laboratories (2).

FDA recognized early that an effective quality assurance program required the full support of local laboratory management and operating personnel. It was also understood that, to a significant degree, each laboratory faced slightly different problems, handled individual workloads, and had to respond to somewhat different types of situations. Therefore each laboratory involved was required to play a primary role in the development of its own program. Headquarters also perceived that it was necessary, for uniformity and comparison purposes, to provide guidelines for the implementation of quality assurance programs, audit procedures, and performance factors for evaluating quality performance.

FDA issued guiding principles that all laboratories were to use to develop their own quality assurance programs. Each laboratory was instructed to develop a plan that could identify quality defects and successes, determine what caused any defect, provide a means for ensuring corrective action, and provide an accounting system. The headquarters guidance plan also included guidelines for the inclusion of performance audit procedures such as oral review of sample worksheets, onsite work reviews, sample accountability checks, instrument performance checks, and so forth. The measurement procedures are reproduced in Appendix D, "FDA Audit Measure Procedures."

These measurement procedures can be helpful in the design of an audit program for any size laboratory, whether it has a permanent or an outside audit team or uses members of the its own staff to assess the laboratory's quality control program. Both FDA field audit personnel and the visiting headquarter's audit teams use the measurement procedures.

Field laboratories were given deadlines for submitting their plans, and periodic reporting dates were established. Overall coordination of the program was retained at headquarters. Quality assurance and audit programs were prepared and are operational on a continuing basis. Reports from the field are evaluated at headquarters, and audits are conducted by a headquarters audit team to assess the quality of the program. Information derived from field reports and from audit team findings is studied to identify trends that might indicate the need for specific local or nationwide training or the need for a change in operational instructions. Any corrective actions that are taken are evaluated and those found to be effective are summarized and communicated to all laboratories in the system.

Recommendations

Every laboratory should include performance and systems audits in its quality assurance program. It is suggested that the following be considered:

1. The use of trained auditors to conduct performance and systems audits with performance audits conducted by laboratory personnel and system audits by an audit unit responsible directly to a level of management above the laboratory level.
2. Performance audits covered on a regular basis for all analysts, including such items as review of written analytical reports, oral review of analytical reports, onsite work reviews, check sample examinations, and intra- and interlaboratory proficiency sample review.
3. System audits be well planned and conducted by a trained audit team in order to minimize the disruption of the laboratory's operations.
4. The site visit be constructive and laboratory personnel be briefed on observations before the team leaves.
5. The report of the audit be prepared as quickly as possible and submitted to the facility for comment.
6. The report highlight deficiencies in operations on a priority basis and make recommendations for corrections.
7. Disagreements on findings be resolved before the report is completed and submitted to management.
8. Follow-up site visits be conducted to see that appropriate corrections were made.

References

(1) Freeberg, F.E. (1980) in *Optimizing Chemical Laboratory Performance Through the Application of Quality Assurance Principles,* F.M. Garfield et al. (Eds), Association of Official Analytical Chemists, Arlington, VA, p. 16

(2) *EDRO Quality Assurance for Field Laboratories* (1978) U.S. Food and Drug Administration, Laboratory Operations Branch, Washington, DC

Chapter 9

DESIGN AND SAFETY OF FACILITIES

Quality analytical work requires facilities that are at least adequate. The suitability of the facility, its location, design, and the internal and external environmental conditions can influence the attitude and response of the staff, the operation and dependability of delicate instruments, and therefore the efficiency and effectiveness of any quality assurance effort.

It is not the purpose of this chapter to specify the design and engineering details of an analytical laboratory, but rather to call attention to certain characteristics of laboratory design and construction that need to be considered when building or remodeling a laboratory. The failure to preplan and then to insist upon important features when dealing with the architects and engineers can lead to poor design, building cost overruns, and an adverse effect on the ability of the laboratory to perform in a quality manner.

The design of a chemical laboratory or the remodeling of an existing one is not a job for amateurs. It requires the expertise of architects, mechanical engineers, electrical engineers, heating and cooling engineers, and other experienced persons who work in these highly specialized fields. But even the experts cannot do the job alone. In addition to information on number of employees and space required, they must receive technical guidance from laboratory personnel on such important matters as needs for specialized laboratories; environmental conditions, including lighting, temperature, and humidity controls; utilities requirements; floor load capacity and stability for certain instruments; electric power needs, including the level of power stabilization and required backup emergency power; work flow patterns; locations, types, and numbers of fume hoods; requirements for special clean rooms or animal facilities; and so forth. New ideas for laboratory design can be obtained from trade journals that annually rate prominent new construction, and when it comes time to actually design a new laboratory, visits to recently constructed installations will provide many ideas on alternative ways to solve various design problems.

An experienced laboratory director will know or sense many of these requirements, having a pretty good idea of space needs, work flow, the kinds of benches and bench tops preferred, the kinds of hoods that are necessary and their best location, how much and what kind of space is needed for storage of chemicals and glassware, and so forth. These thoughts can and should be elaborated, and then reinforced in consultations with senior laboratory scientists who are familiar with the needs of the laboratory, or perhaps with an ad hoc building committee upon which these scientists would serve. The director will need to consult with office and administrative personnel to establish their requirements. If the laboratory is to be part of a much larger installation that will house nontechnical units, the director may become part of a team or committee to assist in the building design or remodeling. In this situation the laboratory director should play a leading or principal role since laboratory requirements are radically different from those of the other units, and many designers and builders may feel that the only difference between a laboratory and an office is the furniture.

Laboratory Design Considerations

The following are some things to consider and some suggestions for laboratory design (1, 2):

- Purpose — The purpose of the laboratory and the kinds of products examined will influence the design. Space that may be adequate for a drug laboratory may not be suitable for a food or a microbiological laboratory.
- People — Obviously, the size of the facility will depend upon, among other things, the number and types of employees and the amount of work space, including desk and record keeping space, they need to do their jobs.
- Location — A central city location, which may be desirable for nonscientific reasons, may entail conformance with stringent building codes that may regulate such matters as the floors on which the laboratory can be located in a multistory office-type structure. Code requirements are less likely to be problems if the facility is located in an industrial park or suburban area, and if the laboratory is in its own building rather than co-located with others. Location must also take into consideration facilities for sending and receiving samples, large shipments of laboratory supplies, and heavy laboratory equipment.
- Workload — Peak and low workloads influence power requirements and mechanical services requirements.
- Expansion — If possible, it is desirable to include in the plans space for laboratory expansion.
- Internal Traffic Patterns — A careful analysis should be made of where analysts, technicians, laboratory workers, and clericals have to move within the laboratory to efficiently perform their duties. This information can be used for planning room sizes, corridor paths, and the configuration of work areas.
- Equipment Location — The location of laboratory equipment will dictate electrical, mechanical, air-conditioning, and humidity requirements. The way in which the equipment is placed can definitely affect efficiency. Sep-

arate instrument rooms, for example, can provide certain efficiencies, but they will require special ventilation to control heat and humidity or other special conditions, and they will not be close at hand for all analysts.

- Vibration and Noise — Many instruments are affected by vibration, including balances, polarographs, and microscopes. Many instruments are also affected by electromagnetic "noise," as one agency finally learned when it realized that an elevator shaft was on the other side of the wall behind their NMR spectrometer. Audible noise can affect the people who work in the laboratory whether they are aware of it or not. Special construction and good planning can avoid these problems. Floor reinforcement may be necessary for especially heavy equipment such as mass spectrometers.

- Receiving Dock — A receiving dock that is remote from the laboratory and the office area will be needed. It is desirable that there be an airlock between the receiving area and the laboratory, as well as other isolation as necessary to prevent vibration.

- Storage Facilities — Separate storage areas for chemicals, glassware, laboratory supplies, gas cylinders, flammable solvents, and hazardous wastes are desirable and usually are required by law. A secure area for gas cylinders is needed where full and empty cylinders can be separated. Special design attention is a must for the storage of flammable solvents with proper ventilation, spill containment floors, explosion-proof electrical fixtures, and the means to contain or extinguish fires.

- Radioactive Areas and Materials — For the design of laboratories that handle radioactive materials, see Mullins' monograph on "Radiochemistry" (3).

- Chemical Fume Hoods and Safety Cabinets — Fume hoods are among the most important features in the laboratory. There are many ways they can be placed in the laboratory, and there are many types of hoods to choose from, depending to some degree on how the laboratory is going to use them. Furthermore, fume hoods can create complex problems by interfering with general laboratory ventilation, heating, and air-conditioning. Safe as well as efficient placement of air-supply and exhaust ducts on the roof of the building can also require difficult decisions. Hood manufacturers can be of assistance in selecting the proper equipment, but good independent engineering services are definitely required before final decisions are made.

- Mechanical and Electrical Systems — The laboratory will require natural gas, compressed air, vacuum, probably two water systems, and electrical power. The major problems with these are their quality, their capacity, and the location of outlets or sources. If a distilled-water system is installed, the delivery system must be noncontaminating. The electrical power must be smoothed and protected against surges and spikes. Backup electrical systems may be needed to protect, for example, frozen or refrigerated sample storage areas and expensive sensitive equipment that could be damaged by power failure. Computer protection must be taken into consideration, since vital information is involved.

- Heating, Cooling, Ventilation, and Illumination — Air-conditioning and lighting requirements are quite different for various areas in the laboratory as well as for the office areas. These conditions affect the ability of people to operate efficiently, although it is probably the demonstrable effect on the performance of chromatographs and volumetric apparatus that most convinces laboratory designers. The need for windows and daylight in the laboratory must be considered and resolved.

- Special Laboratory Facilities — Special laboratory facilities may be necessary. For example, special provisions for contamination control are required if analyses will be performed for trace residues of a chemical that is ubiquitous in the environment, such as lead, or for chemicals where low levels must be measured, such as chlorinated dibenzodioxins. Also, special safety design provisions are necessary to protect the health of those conducting investigations on microbiological agents or highly toxic substances such as antineoplastic drugs. If live animals are to be used, special housing facilities will be necessary, preferably away from other areas of the laboratory.

- Fire and Explosion Hazards — Detectors, alarms, emergency lighting, fire extinguishers, fire containment design, fire doors, fire showers, and automatic sprinkler systems must be provided. Also, two means of egress from each laboratory in the event of an emergency and fireproof construction materials are necessary in some areas.

- Wall, Ceilings, and Floor — Nonporous and easily cleaned materials are required for walls, ceilings, and floors. Floors should be skid resistant and sturdy so that, for example, tiles are not going to curl into hazards under the influence of normal laboratory operations. Walls and ceilings should be light and cheerful colors that will not distort lighting.

- Laboratory Furniture — Laboratory furniture is preferably of metal construction with acid- and solvent-resistant work surfaces. Storage cabinets should be built of materials that will withstand the corrosive effects of materials that may be stored in them. All areas where utilities may pass through must be available for repairs, including internal areas of laboratory benches. Desks and other furniture should be sufficient for the need, but comfortable enough for long hours of sedentary work. Exposed overhead piping and flat-top wall cabinets are to be avoided. Laboratory furniture manufacturers can suggest designs and specifications.

- Codes and Standards — Although these are problems for the designer and the engineers, it is useful and helpful for the laboratory building committee to have some knowledge of the various code requirements. Pertinent requirements for laboratories are stated in the U.S. Occupational Safety and Health Act (OSHA) and regulations, and helpful information on safety is

available in the voluntary standards of the National Fire Protection Association (NFPA).
- Telephone System — Telephone companies can help design an efficient and minimum-cost system, but a decision will have to be made by laboratory management on the preferred configuration. For example, there are many advantages to having a public address system, both for paging people who are being called and for making general as well as safety-related announcements.
- Computers — Computers require certain environmental conditions but none more stringent than required by other instruments in the laboratory. An important consideration is where to install computer cabling, which can feed outlet plugs for connecting computer equipment. If a central computer room for a mainframe or minicomputer is to be installed, expertise can be obtained from computer companies.
- Office Design — At least three types of areas must be considered here: the work area for the analysts to do their calculations and other paperwork, the first-line supervisors' offices, and offices for second-line and top management. In all cases, good lighting, good but not breezy or noisy ventilation, and adequate flat surface are important. Take into account the fact that people these days work with a computer, or at least a terminal, on their desks, and that in addition to computers there are reference books, manuals, and plenty of paperwork that needs to be stored, preferably out of sight.
- Special Rooms — Provisions must be made for such special areas as a room for lunches and breaks; a conference room that will accommodate the entire staff and, in addition, smaller rooms for seminars or training meetings; a library (which can be combined with a conference room), sample preparation and mixing rooms; sample storage rooms, including refrigerated and freezer storage; a room for glassware washing and drying; refuse disposal rooms, and glassware and chemical storage areas.

The reliability of analytical results depends upon the skill of the analysts, their supervision and motivation, the use of well-tested methods, and properly maintained analytical instruments. The configuration and design of the laboratory, however, as well as its location and the quality of the laboratory environment, can also significantly affect the performance of the staff as well as the instruments, and therefore the success of the laboratory's operation.

Laboratory Safety

By anyone's definition of safety, a chemical laboratory is not a safe place. In actuality, however, because analysts generally recognize the hazards, major accidents can be avoided and minor ones reduced to a tolerable minimum. Key to this situation, of course, is effective employee training; appropriate monitoring of potentially dangerous equipment, areas, and practices; and the inclusion of safety considerations in the design of the facility. One safety manual states, "Basically, safety is affirmative action, based on knowledge of the hazards and other properties of materials" (4).

Chemistry students are generally given a perfunctory lesson on safety at their first laboratory session. This comes at a time when they may be ill-prepared to understand the importance of the information. Stress is placed on fires and flammable solvents, chemical and other types of burns, use of eye protection, and injuries that can result from carelessness in the use of glassware and other laboratory materials. In an organic chemistry course, attention is generally given to hazards from spills and the inhalation of hazardous and toxic chemicals. The student, however, will pick up safety practices from experience, or from near experience, with the laboratory routine, as well as from the experiences of others. This information may be reinforced from time to time with special safety seminars at school and in the workplace.

The hazards are ever present, however, and accident avoidance as well as employee understanding of proper accident handling must be given continual attention. A laboratory safety committee, involving top management as well as bench scientists, is most essential for giving appropriate prominence to safe laboratory practices. The committee will be a focal point for receiving comments from employees concerning unsafe situations, addressing unsafe situations, and providing periodic training to the staff on how to avoid or how to deal with particularly dangerous situations.

There are many publications on laboratory safety and working with hazardous chemicals. Although there is much duplication in this material, most publications seem to take a different or unique approach, or stress one or more features of the safety problem. Several texts should be stocked in the library of every scientific laboratory and used for instruction at safety seminars and for reference purposes.

National chemical societies and associations stress safety programs and, in their journals, publish articles or reports on unusual laboratory fires, accidents, or explosions. The American Chemical Society has an active Committee on Chemical Safety as does the Association of Official Analytical Chemists. In its *Official Methods of Analysis,* the AOAC includes a chapter on "Laboratory Safety" that contains general information on the potential hazards of laboratory equipment; general safety techniques and practices; safe handling of acids, alkalies, and organic solvents; and safe handling of special chemical hazards.

Each laboratory must have a safety program and give attention to safety as part of the orientation of new employees. At a minimum, a person must be designated as the safety officer or coordinator, and an active safety committee is highly recommended as a means to get analysts involved in thinking about safety. A laboratory of any size will need a safety manual, whether of its own design or adopted from another organization. A large organization with several laboratories in different locations should consider decentralizing its program so that each laboratory will have one that suits its particular situation, but which will establish general program requirements. All must be meticulous in reporting all accidents and employee injuries, both for purposes of establishing and monitoring their safety record and for protecting themselves from lawsuits on the basis of negligence.

Safety In Facility Design

Each laboratory must be designed to meet applicable safety codes and standards as required by the governmental jurisdictions in which they are located, with special attention to fire and explosion hazards. The following is a suggested list of safety design features for consideration (5, 6):

- Early detection and fire alarm systems and automatic fire-extinguishing systems for storage rooms, solvent rooms, and interior of fume hoods
- Hood exhaust blowers of spark-proof construction and explosion-proof motors if mounted in the air stream
- Fire emergency circuits that can shut down building ventilation and hood blowers
- Services to hoods controlled by valves & switches outside of the hoods
- Storage cabinets for flammable chemicals that meet NFPA Standards
- Fire safety doors in suitable locations and two exits from each laboratory
- All emergency exits lighted or with luminescent exit signs, and building exit doors opening outward and equipped with push bars
- Emergency lighting throughout the laboratory areas to provide sufficient lighting in case of a power failure
- Explosion-proof electrical outlets where combustible solvents may be a problem
- All apparatus grounded and extension cords prohibited
- Lead shielding where gamma-ray emitters are used and special construction of work surfaces to prevent contamination with radioactive materials
- Safety showers and eyewash stations placed at strategic locations and checked frequently for proper functioning.

Safety Equipment

All laboratories provide their scientists with protective clothing and other protective items. Failure to provide these items, or, if provided, failure to instruct employees on their use and to insist upon their use have caused unnecessary injuries. Laboratory coats or aprons, gloves, and eye protection devices must be readily available and their use strictly enforced. For special operations, rubber aprons, heat-resistant gloves, face masks or goggles, and hearing protection should be provided and, again, their use by the staff under appropriate circumstances enforced. For special operations, heavy rubber shoe coverings, caps, steel-toed shoes, straps to tie gas cylinders, warning signs, and flashlights must be available, while gas masks, respirators, first-aid kits, decontamination materials ("spill kits"), fire extinguishers, sand buckets, and fire blankets should be conveniently located and their locations clearly marked so that they can be found quickly in an emergency.

Hazardous Materials

Chemical handling in the laboratory is so commonplace that laboratory directors and analysts tend to overlook the legal requirements and often the safety problems that are associated with their use. The Occupational Safety and Health Administration (OSHA) in the Department of Labor issued a final rule for occupational exposure to hazardous chemicals in laboratories under Title 25 of the *Code of Federal Regulations,* Part 1910 (7). This rule applies to all laboratories that use hazardous chemicals. A laboratory is defined as "a facility where the laboratory use of hazardous chemicals occurs. It is a place where relatively small amounts of hazardous chemicals are used on a nonproduction basis." A hazardous chemical is defined as "a chemical for which there is statistically significant evidence, based on at least one study conducted in accordance with established scientific principles, that acute or chronic health effects may occur in exposed employees." The term health hazard is defined as including "chemicals which are carcinogens, toxic or highly-toxic agents, reproductive toxins, irritants, corrosives, sensitizers, hepatotoxins, nephrotoxins, neurotoxins, agents that act on the hematopoietic systems, and agents which damage the lungs, skin, eyes, or mucous membranes." The effective date of the rule was May 1, 1990.

There is also a requirement that employers must complete an appropriate *Chemical Hygiene Plan* and carry out its provisions by January 31, 1991. This plan is "a written program developed and implemented by the employer which sets forth procedures, equipment, personal protective equipment and work practices that [1] are capable of protecting employees from the health hazards presented by hazardous chemicals used in that particular workplace and [2] meets the requirements of paragraph [e] of this section."

Appendix A of the *Federal Register* announcement (Section 1910.1450), which is nonmandatory, is offered as a guide for employers in the development of an appropriate laboratory Chemical Hygiene Plan. The suggested guidelines are extracted from a 1981 publication of the National Research Council, *Prudent Practices for Handling Hazardous Chemicals in Laboratories* (8) (referred to below simply as *Prudent Practices*), which can be obtained from the address given in the reference. The final rule itself can be obtained from the Office of Information and Consumer Affairs, OSHA, 200 Constitution Avenue, NW, Room N3649, Washington, DC 20210.

Prudent Practices suggests that responsibilities for chemical hygiene rest at all levels, including the chief executive officer, supervisor of the department, chemical hygiene officers, laboratory supervisors, and laboratory workers. The text then covers such areas as laboratory facility design, chemical procurement, environmental maintenance, medical programs, protective apparel, safety equipment, and so forth.

Appendix B of the *Federal Register* announcement (Section 1910.1450) provides references to assist the employer in the development of the Chemical Hygiene Plan.

The final rule itself, which is lengthy, has an important section on employee information and training. It requires that the employer provide employees with information and training to ensure that they are apprised of the hazards of chemicals present in their work area. The information includes contents of the standard or final rule; location and availability of the Chemical Hygiene Plan; and information permissible exposure limits for OSHA-related substances, signs and symptoms associated with hazardous chemicals used in the laboratory, and the availability and location of known reference materials on the hazards, safe handling, storage, and disposal of hazardous chemicals. This information is available on the Material Safety Data Sheet that all chemical suppliers are required to furnish with materials they sell. These sheets must be available in the laboratory for the inspection by any employee who might be exposed to a chemical.

The rule also requires, as part of the Chemical Hygiene Plan, the formulation of and strict attention to a program for the disposal of waste laboratory chemicals. The National Academy of Sciences has also published a text, *Prudent Practices for Disposal of Chemicals from Laboratories,* that provides guidance that is within the rules established by the U.S. Environmental Protection Agency (9). This book can be helpful in designing the Chemical Hygiene Plan.

Emergency Control Procedures

Each laboratory needs a written plan to cover fires, fire fighting, and evacuation of the premises if an emergency occurs. The fire response plan should cover both "fight" and "flight" situations. The decision whether to fight a fire or evacuate a building is difficult. In some instances, both actions may be indicated. Small bench-top fires are among the most common laboratory emergencies, and in most cases these fires can be extinguished without summoning the fire department or evacuating the building. An individual confronted with a fire, however, should make certain that another person is aware of the fire and is summoning assistance before attempting to extinguish it single-handed. Any fire, no matter how apparently minor, may spread with unexpected rapidity, thus the importance of making certain that help is on its way before attempting any heroics. When notification has been confirmed, the immediate action against the fire can consist of use of a fire blanket or a suitable fire extinguisher. Meanwhile, the person in charge, a supervisor, or the safety officer should review the situation and decide whether to continue fighting the fire or evacuate the building and summon the fire department.

All personnel must receive instructions on the fire response plan and be trained to perform assigned responsibilities. The plan, including building evacuation maps indicating appropriate exits, must be posted at several conspicuous locations. Emergency telephone numbers for doctors, hospitals, ambulances, fire departments, and others must be posted near each telephone and kept up to date. Periodic contact with these service facilities is desirable so that they are kept aware of the location of the laboratory and the types of service they may be expected to furnish. It is advisable to conduct a fire drill, including total building evacuation, at least twice a year, with identified relocation areas and each drill evaluated in writing and corrections made in drill procedures as indicated.

Safety equipment requires regular checking to ascertain that items are in their proper places and functioning properly.

Each laboratory should have at least a number of individuals trained in first-aid and in cardiopulmonary resuscitation by the Red Cross, Civil Defense, or other recognized organization. The number of trained people should be sufficient to ensure the presence of someone trained at all times.

Good housekeeping is important for reducing laboratory risks and hazards, and such requirements belong in the safety plan. Accident reports are required for all accidents in which there is injury or potential injury. These reports are useful for insurance reasons, possible litigation, and especially for management review to improve preventive measures and employee safety education.

Personal Habits and Safe Operating Practices

All laboratories seem to insist upon certain safety rules for personnel, but the way this is handled varies widely and there is a general problem with enforcement. It is strongly recommended that the requirements be developed by management and representatives of the laboratory staff, put in writing in the safety program manual, and that they be reviewed from time to time as part of the training initiative in the safety program. The following precautionary rules or practices, taken from several sources, are suggested:

- Food, candy, gum, and beverages are to be stored and consumed *outside* the laboratory.
- No smoking is to be permitted in the laboratory or anywhere else that it might endanger employee safety or health. Areas for smoking are to be designated.
- Testing of samples or chemicals by taste must be forbidden, and odors should only be checked with care.
- Laboratory coats, gloves, masks, or other apparel must not be worn when leaving the laboratory for a public area or a place where food is being consumed.
- Hand washing is required after removing protective gloves and after returning to the laboratory from rest rooms, or from other outside areas.
- Personal items, such as coats, hats, umbrellas, and purses, are to be stored in lockers and not worn or carried through the laboratory.
- "Horseplay" is absolutely forbidden in the laboratory.
- Persons with long hair are required to tie it in back or cover their heads with some form of cap.
- Beards should be discouraged or cut short.
- Water for drinking must be located outside the laboratory, or foot-operated fountains should be provided.
- Desk tops must be kept free of clutter and unnecessary paper, chemicals, and equipment.
- Aerosols are to be used in hoods and not at bench areas.

- Cleansing tissue rather than handkerchiefs should be used when necessary for personal purposes.
- Working *alone* outside of normal working hours is prohibited.
- The use of pipet filling bulbs is required for all pipet use.
- Contact lenses may not be worn in the laboratory.
- Protective safety glasses are required at all times or, at the very least, in any potentially hazardous situation.
- Face shields are required when potential spill, splatter, or impact conditions may occur.
- Appropriate warning signs must be used when hazardous conditions may occur.
- All chemical storage containers must be labeled; unlabeled bottles are automatically discarded.
- Separate, covered, metal waste containers are to be provided for paper and broken glassware, and special arrangements are made for the disposal of solvents and other hazardous wastes.
- Used glassware must be emptied of solutions and solvents and rinsed with water before being released for regular cleaning, and if special instructions for cleaning are necessary, cleanup personnel must be informed.
- Chipped and cracked glassware must be destroyed.
- All laboratory experiments are to be reviewed for possible safety problems.
- Safety shields are required around high-vacuum or high-pressure reactions.
- Ongoing reactions must be attended at all times.
- Gas cylinders are to be secured before protecting caps are removed.
- Laboratory visitors should be restricted. If persons are allowed in the laboratory, they must be accompanied by a member of the staff and provided with eye and head protection, as necessary.

There are many other personal safety requirements that can and should be part of the laboratory's safe practice requirements.

Wilcox et al. (*10*) observe that

> Good safety practices are necessary to prevent time lost from accidents and damage to equipment and facilities. Concern for the personal welfare of the employees is equally important. From a quality assurance standpoint, also, accident prevention is vital. In general, maintaining quality performance depends on a smooth work flow. The sudden loss of an employee due to injury may upset the work flow sufficiently to compromise the quality of the day's testing. Thus, any laboratory accident should be followed by an immediate assessment by the director or supervisor that results from the laboratory will remain accurate.

Recommendations

Before a new laboratory facility is built or an old one remodeled, the following suggestions and recommendations should be considered.
1. A competent architect and engineers experienced in laboratory design be employed.
2. An ad hoc building committee be appointed to advise the designer and engineers on special laboratory requirements.
3. Expansion potential be built into the design.
4. Special attention be given to safety design features that are essential to safety in laboratory operations.

As part of the safety plan of the laboratory, the following should be considered:
1. Development of a written safety plan and Chemical Hygiene Plan.
2. When necessary, the laboratory be modified to incorporate safety features to protect personnel.
3. Appointment of a safety officer or safety committee with responsibility for safety.
4. Indoctrination of personnel in the safety aspects of the plan and the Chemical Hygiene Plan, and specific assignments made.
5. Conduct of fire and evacuation drills on a scheduled basis.
6. Placement of various types of safety equipment at strategic locations and regular checking to ascertain that the items are in place and working.
7. Establishment of good housekeeping practices.
8. Development of personal hygiene habits and safe practices with which personnel are required to comply.
9. Periodic conduct of safety inspections and fire drills with follow-up to ensure that problems are corrected.

References

(1) Rappaport, A.E., et al. (1977) in *Quality Assurance Practices for Health Laboratories,* S.L. Inhorn (Ed.), American Public Health Association, Washington, DC, p. 1

(2) Bloom, H.M. (1980) in *Optimizing Chemical Laboratory Performance Through the Application of Quality Assurance Principles,* F.M. Garfield et al. (Eds), Association of Official Analytical Chemists, Arlington, VA, p. 29

(3) Mullins, J.M., et al. (1977) in *Quality Assurance Practices for Health Laboratories,* S.L. Inhorn (Ed.), American Public Health Association, Washington, DC, p. 1007

(4) Committee on Chemical Safety (1979) *Safety in Academic Chemistry Laboratories*, 3rd Ed., American Chemical Society, Washington, DC, p. iii

(5) Rappaport, A.E., et al. (1977) in *Quality Assurance Practices for Health Laboratories,* S.L. Inhorn (Ed.), American Public Health Association, Washington, DC, p. 202

(6) Bloom, H.M. (1980) in *Optimizing Chemical Laboratory Performance Through the Application of Quality Assurance Principles,* F.M. Garfield et al. (Eds), Association of Official Analytical Chemists, Arlington, VA, p. 30

(7) *Fed. Regist.* (Jan. 31, 1990) 55(21), 3300–3335

(8) National Research Council (1981) *Prudent Practices for Handling Hazardous Chemicals in Laboratories,* National Academy Press, Washington, DC

(9) *Prudent Practices for Disposal of Chemicals from Laboratories* (1983) National Academy Press, Washington, DC

(10) Wilcox, K.R., et al. (1977) in *Quality Assurance Practices for Health Laboratories,* S.L. Inhorn (Ed.), American Public Health Association, Washington, DC, p. 81

Chapter 10
LABORATORY ACCREDITATION

Government agencies and many professional and private organizations have recognized the need to evaluate and upgrade the performance of laboratories that operate in various disciplines, or to periodically check their ability to test specific products for conformance to specified standards. This need has led to a proliferation of so-called accreditation programs that are being widely used to certify laboratories and provide some assurance of their ability to provide accurate test or inspection results. Accreditation is usually specific for tests of the systems, products, components, or materials for which the laboratory claims proficiency.

A study by the U.S. Department of Commerce under its National Voluntary Laboratory Accreditation Program identified and examined 26 programs that are directly operated or sponsored by the federal government, 20 state and local programs, 24 professional and trade association programs, and over 50 that are handled by private organizations (*1*). There are an estimated 50,000 laboratories, ranging from one-person test stations to complete multinational, multidisciplined commercial laboratories, that are approved or accredited in the United States under the privately operated programs alone.

Most of these laboratories deal with mechanical and electrical testing, heat and temperature measurement, optics and photography, and other nondestructive physical testing. Only about a dozen of these accreditation programs pertain to chemical analytical laboratories.

The fact that so many accreditation programs exist seems to confirm that there is a need for them. Private-sector and governmental purchase of components and subsystems appears to be the largest use of the laboratory accreditation systems.

There is also considerable activity in the development of laboratory accreditation programs in conjunction with international trade negotiations. The Agreement on Technical Barriers to Trade, drawn up within the framework of the General Agreement on Tariffs and Trade (GATT), has been signed by more than 30 governments and by the European Economic Community (EEC). These countries have accepted internationally binding obligations to reduce unnecessary obstacles to trade that arise from technical regulations, standards testing arrangements, and certification systems. But the need for accreditation systems is clearly acknowledged.

The International Laboratory Accreditation Conference (ILAC) has identified the technical conditions that enable governments to place confidence in the competence and reliability of testing bodies (2). The United States initiated the first ILAC meeting in Copenhagen in 1977, and meetings are held biannually. The Conference is a coordinated effort of more than 40 countries to examine how laboratory accreditation systems in the various participating countries can be used in bilateral or multilateral contracts or agreements.

Although committee arrangements change from time to time, ILAC has been concerned with the following: (a) legal aspects of the international reciprocity of test data, (b) the format for an *International Directory of Laboratory Accreditation Systems and Other Schemes for Assessment of Laboratories* (the 3rd edition of the directory was published in 1985), and (c) definitions and common criteria for national accreditation systems.

Future efforts of ILAC are expected to focus on technical matters as well as on the implementation of agreements between nations for the mutual acceptance of test data. The latter effort is being undertaken cooperatively within the United Nations Economic Commission for Europe to include all eastern European countries as well.

Institutional accreditation is not a new phenomenon, although it is relatively new for some areas of laboratory operations. Hospitals and colleges have been subject to accreditation for many years. Hospitals, for example, including their laboratories, must be evaluated regularly according to criteria established by the medical community.

As mentioned earlier, clinical laboratories have been subjected to regulation under the U.S. Clinical Laboratories Improvement Act of 1967 (CLIA) (3). This act provides, in part, that any laboratory that solicits or accepts, in interstate commerce, directly or indirectly, any specimens for laboratory examination must obtain a license issued by the Secretary of Health and Human Services.

Several professional organizations, too, have established programs for accreditation of chemical laboratories. These include the College of American Pathologists, the American Society of Cytology, the American Association of Blood Banks, and the American Industrial Hygiene Association. The standards used by these organizations are similar to those required by government agencies. The main difference is that government licensure is mandatory whereas association programs are voluntary. Several of these associations conduct proficiency test programs.

Approaches to Accreditation

There are two basic approaches to accreditation of laboratories. The first approach accredits laboratories to test specific products in conformance with relevant standards, using test methods pertinent to the product (4). These are product-focused programs. The second approach accredits laboratories to conduct tests in broad areas or groups of products. These are discipline-focused programs.

For example, in the United States the National Voluntary Laboratory Accreditation Program (NVLAP) (5) of the U.S. Department of Commerce is product oriented, and the system of the American Association for Laboratory Accreditation (AALA)

(6) is discipline- or laboratory-oriented. In the latter program, each laboratory may be accredited for more than one discipline and for more than one subdiscipline within each discipline. This approach permits laboratories to be evaluated for the entire range of their capabilities rather than requiring a separate process for each test or product.

Objectives of Laboratory Accreditation Systems

Laboratory accreditation systems, designed to meet particular needs, vary in format and substance (7). Specific objectives that laboratory accreditation systems tend to include are:

- Ensure the validity of test data
- Promote the acceptance of test data by users of laboratory services so that data produced by one laboratory can be accepted by another without further tests
- Facilitate international trade through the acceptance of test data from accredited laboratories
- Make more efficient use of testing facilities within a country by coordinating existing capability
- Add to the credibility of more laboratories
- Give additional status to competent laboratories
- Promote good testing practices
- Improve testing methods by providing feedback to standards-producing bodies on the adequacy of test methods
- Provide technical and other information to accredited laboratories

Almost all of these systems aim to improve laboratory operations through the use of quality assurance programs. Emphasis is placed on the correction of deficiencies rather than on rejection or denial of accreditation.

Accreditation Criteria

General criteria for laboratory accreditation have been developed by several associations. The American Society for Testing and Materials (ASTM) has published a standard that is titled "Standard Recommended Practice for Generic Criteria for Use in the Evaluation of Testing and Inspection Agencies" (ASTM-E548), which suggests generic criteria intended for use by accrediting authorities for qualifying and accrediting testing agencies (8). The criteria deal with organization, staff, facilities, equipment, and quality control. Attention is also being given to a model for an accreditation system consisting of an accrediting authority, accreditation criteria, and an evaluation and monitoring program. There is a move toward defining areas of testing or discipline in terms of test method groupings, procedures, or techniques.

Protocols have also been developed by ISO (International Organization for Standardization) as ISO Guide 25, "Guidelines for Assessing the Technical Competence of Testing Laboratories" (9); and by ANSI (American National Standards

Institute) as "American National Standard for Certification-Third Party Certification Program" (*10*).

In 1982–83 the Association of Official Analytical Chemists, under contract with the Ministry of National Health and Welfare Canada, conducted a study "To Develop and Test an Analytical Laboratory Accreditation Program for Food and/or Drug Laboratories."

As part of the study the AOAC, working with a select committee of scientists with quality assurance and laboratory management experience, developed a series of protocols to implement an accreditation program. This included accreditation procedures, accreditation criteria, an organizational structure for administration, guidelines and application for accreditation, a training program for auditors, and a laboratory inspection checklist. The accreditation criteria developed for judging a laboratory's competence are based largely on the laboratory's quality assurance program and cover such elements as organization, human resources, material resources, quality system, preventive maintenance for equipment, quality of supplies, sample handling and record system, test records, test methods and procedures, validation of performance, and deficiency correction.

For information purposes, the AOAC accreditation criteria are reproduced in Appendix F. As a minimum, the criteria can be used by laboratories in a self-evaluation program against which their quality assurance systems can be measured and compared. For laboratories that do not have a documented quality assurance program, the criteria can be used as a guide for the preparation and initiation of one, or in anticipation of seeking accreditation.

National Accreditation Programs

Several laboratory accreditation programs outside of the United States are authorized and supported, at least in part, by national governments. Australia, New Zealand, the United Kingdom, and Canada, for example, operate such programs. The Australian program, known as the National Association of Testing Authorities (NATA), developed from a 1946 government directive to establish a national testing laboratory accreditation system (*11*). The system was designed to provide a national testing service that would satisfy the needs of the government, industry, and commerce; to operate uniformly throughout the country; and to assess the competence of laboratories through comprehensive examinations of their testing operations by independent experts. The assessment is to be based on true peer-group review, with emphasis on management evaluations.

NATA accreditation is granted for specific tests, which are grouped into fields of testing based more or less on scientific disciplines. There are nine fields of testing: (a) acoustic and vibration measurement, (b) biological testing, (c) chemical testing, (d) electrical testing, (e) heat and temperature measurement, (f) mechanical testing, (g) nondestructive testing, (h) metrology, and (i) optics and photometry. The referenced article covers the program and provides details on structure, funding, accreditation criteria, assessors, applications for accreditation, assessment procedures, and a report of the assessment.

The assessment procedures are very similar to those suggested in Chapter 8, "Audit Procedures," in this book. They include attention to the following: staff structure; management attitudes; the facility; organization and management; procedures for the collection, transport, handling, and storage of test samples; procedures for the receipt of samples by the laboratory, including their logging and identification; criteria for rejecting samples; selection or specification of tests to be applied; allocation of testing work to staff members; instruction and supervision of staff; laboratory quality control programs; procedures for recording test data; procedures for processing test data and checking calculations; procedures for reporting, with special emphasis on statements of accuracy or precision; and security systems.

The Standards Council of Canada (SCC), which is the accrediting authority for the national voluntary accreditation program for testing organizations in Canada, was created by an act of Parliament in 1970 (12). It consists of 16 federal and provincial officials and 40 representatives from private organizations that are interested in standardization. The prime objective is to foster and promote voluntary standardization. Under the act, SCC may "accredit in accordance with criteria and procedures adopted by the Council, organizations in Canada engaged in standards formulation, testing, and certification."

Specialized expertise is furnished to the council by various advisory committees. The Advisory Committee on Certification and Testing (ACCT) provides advice on those topics.

The Council published "Accreditation Criteria and Requirements for Testing Organizations" (CAN-P-4A), which outlines the general criteria and requirements for testing organizations accredited by the Standards Council of Canada (13). The Council approved "Guidelines for Preparing a Quality Manual for Testing Organizations" (CAN-P-1511), which is intended "to assist a testing organization to set forth, in a systematic way, the measures by which it implements its internal Quality System" (14). It is also intended as a guide "for the preparation of procedures by which a testing organization describes those measures which it intends to employ to give confidence that its work meets its quality objective." This is an interesting document that provides, under each heading and subheading, information critical to portraying an accurate picture of how the quality system is intended to work.

From the foregoing, and from a review of other accreditation programs, it can be seen that these programs have very similar objectives and purposes as quality assurance programs. In accreditation, the approving body is external to the organization that is seeking accreditation. In quality assurance, the program and the assessment are carried on in-house. The certification or accreditation of laboratories largely depends on the quality assurance program. Studies have shown that a good quality assurance system, with proper documentation, provides the basics for accreditation.

Good Laboratory Practice

In December 1978, the U.S. Food and Drug Administration issued final regulations regarding good laboratory practices in the conduct of nonclinical laboratory studies (*15*). According to the *Federal Register* announcement,

> The action is based on investigatory findings by the agency that some studies submitted in support of the safety of regulated products have not been conducted in accord with acceptable practice, and that accordingly data from such studies have not always been of a quality and integrity to assure product safety in accord with the Federal Food, Drug and Cosmetic Act and other applicable laws. Conformity with these rules is intended to assure the high-quality of nonclinical laboratory testing required to evaluate the safety of regulated products.

The regulations spell out what is expected of an organization that engages in studies to support the safety of regulated products. The criteria are similar to the requirements of any good quality assurance program. As a matter of fact, FDA requirements have fostered interest in laboratory quality assurance and the interest of some private and governmental agencies in accreditation programs. The regulations provide for inspection of the testing facility by an authorized employee of FDA. The FDA will not consider a study to be in support of an application if the facility will not permit an inspection. On September 4, 1987, the FDA issued a final rule that amends these regulations to clarify them and reduce the burden on testing facilities (*16*).

FDA has also promulgated regulations for Good Manufacturing Practices for Human and Veterinary Drugs that contain requirements for laboratory controls under Section 211.160 (*17*). This section states, in part, "laboratory controls shall include the establishment of scientifically sound and appropriate specifications, standards, sampling plans, and test procedures designed to assure that components, drugs product containers, closures, in-process materials, labeling, and drug products conform to appropriate standards of identity, strength, and purity." The regulations call for a quality control unit, and they spell out what the laboratory controls must include.

The U.S. Environmental Protection Agency (EPA), too, has issued regulations for good laboratory practice standards for toxic substances and for pesticides (*18, 19*). These regulations parallel those of the FDA, but are more specific in certain respects. In March 1978, EPA entered into an Interagency Agreement with FDA that formalized the cooperative efforts of the two agencies to establish a coordinated quality assurance program for their various activities (see *Federal Register*, April 4, 1978, p. 14124). The two agencies work closely in developing and conducting their programs.

The Organization for Economic Cooperation and Development (OECD), in which some 20 countries and a half-dozen international organizations participate, in 1982 issued a final report of an expert group, titled *Good Laboratory Practice in the Testing of Chemicals* (*20*). OECD was set up under a convention signed in Paris on December 14, 1960, with several purposes, including one "to contribute to the expansion of world trade on a multilateral, nondiscriminatory basis in accordance with international obligations." The report contains four chapters, two of which are of special interest since they relate to laboratory quality assurance: "Principles of Good Laboratory Practice," and "Guidelines for National GLP Inspections and Study Audits." The chapter on principles of good laboratory practices provides guidance to test facilities to promote the development of quality control data in the testing of regulated chemicals, including industrial chemicals, pharmaceuticals, and pesticides. The chapter on inspections and study audits discusses an inspection program that is intended to provide guidance in ascertaining proper study conduct in order to ensure that data produced are of adequate quality to permit hazard assessment.

Certification, Registration, or Licensing of Chemists

Because personnel qualifications are so important to a quality assurance program, accreditation program, or good laboratory practices regulations, some attention must be directed toward the certification, registration, or licensing of chemists. An interesting article on the subject, "Do Chemists Need Added Credentials" was published in *Chemical and Engineering News* (*21*). According to the article, terms are a source of continuing confusion. The term "certification," as ordinarily used, refers to a voluntary process carried out by a nongovernment organization, whereas "registration" and "licensing" refer to procedures performed by government agencies. Registration is usually optional, but licensing is required to practice certain functions. There are arguments in favor of certification or licensing and arguments against. The best argument for certification is that it tends to improve professional competence in that it identifies persons who have special competence in their fields; thus it helps employers select individuals who meet at least minimum standards of performance. The argument against certification is also strong: the record of a person's academic and job experience, as well as evaluations of professional performance by present and former employers, should provide enough information to make certification unnecessary. The assumption that a person who is licensed or certified is more qualified than one who is not is of questionable validity.

The strongest argument in favor of licensing is that it would increase prestige and public recognition by placing chemists in a professional status with physicians and certain engineers, who must be licensed. Licensing would tend to keep unqualified persons out of the field. A strong argument against licensing is that chemists do not have to sign legal documents, therefore there seems to be little reason for requiring the time and costs involved in establishing and administering a licensing program.

The American Chemical Society and the American Institute of Chemical Engineers do not have programs for certifying members. The American Institute of Chemists does have a certification and a recertification program.

Conclusions

From the foregoing, a few conclusions can be drawn:

- Quality assurance programs and laboratory accreditation both have as their primary goals improvements in laboratory operations and confidence in the results.
- If a laboratory does not have a good quality assurance program, it will have great difficulty in obtaining accreditation.
- The key to systems of quality assurance and accreditation is documentation.
- No laboratory can operate successfully without a quality assurance program.

References

(1) *Principle Aspects of U.S. Laboratory Accreditation Systems* (1980) U.S. Department of Commerce, Washington, DC

(2) *International Directory of Laboratory Accreditation Systems and Other Schemes for Assessment of Testing Laboratories*, 3rd Ed. (1985) International Laboratory Accreditation Conference, Paris, France

(3) "Clinical Laboratory Improvement Act of 1967," Part F, Title III, Public Service Health Act, Section 353

(4) Locke, J.W. (1980) in *Testing Laboratory Performance: Evaluation and Accreditation*, NBS Publication 591, G.A. Berman (Ed.), National Institute of Standards and Technology, Gaithersburg, MD, p. 6

(5) Unger, P.S. (1980) *ASTM Standardization News* 8(11), 18

(6) Amorosi, R.J. (1980) *ASTM Standardization News* 8(11), 21

(7) Locke, J.W. (1982) *Laboratory Accreditation: Future Directions in the United States*, NBS Publication 632, National Institute of Standards and Technology, Gaithersburg, MD, p. 43

(8) "Standard Recommended Practice for Generic Criteria for Use in Evaluation of Testing and/or Inspection Agencies," ASTM E548 (1979) American Society for Testing and Materials, Philadelphia, PA

(9) "International Standards Organization, Guide 25" (1982) American National Standards Institute, New York, NY

(10) "American National Standard for Certification — Third Party Certification Program," ANSI Z34.1 (1982) American National Standards Institute, New York, NY

(11) Monaghan, H.F. (1980) *ASTM Standardization News* 8(11), 10

(12) MacNintch, R.E. (1980) in *Testing Laboratory Performance: Evaluation and Accreditation*, NBS Publication 591, G.A. Berman (Ed.), National Institute of Standards and Technology, Gaithersburg, MD, p. 88

(13) "Accreditation Criteria and Requirements for Testing Organizations" (1984) Standards Council of Canada, Ottawa, Ontario

(14) "Guidelines for Preparing a Quality Manual for Testing Organizations," (1987) Standards Council of Canada, Ottawa, Ontario

(15) *Fed. Regist.* (Dec. 22, 1978) "Nonclinical Laboratory Studies, Good Laboratory Practice Regulations," p. 60013 (21 CFR Part 58)

(16) *Fed. Regist.* (Sept. 24, 1987) "Good Laboratory Practice Regulations," p. 33768 (21 CFR Part 58)

(17) *Code of Federal Regulations,* Title 21, Part 210, "Current Good Manufacturing Practices in Manufacturing, Processing, or Holding Drugs," and Part 211, "Current Good Manufacturing Practices for Finished Pharmaceuticals"

(18) *Fed. Regist.* (Nov. 29, 1983) "Toxic Substances Control, Good Laboratory Practice Standards," p. 53922 (40 CFR Part 792); and "Pesticide Program, Good Laboratory Practice Standards," p. 53946 (40 CFR Part 160)

(19) *Fed. Regist.* (Dec. 28, 1987) "Federal Insecticide, Fungicide, and Rodenticide Act (FIFRA) and Toxic Substances Control Act (TSCA), Good Laboratory Practices Standards," p. 48920 (40 CFR Parts 160 and 792)

(20) *Good Laboratory Practice in the Testing of Chemicals* (1982) Organization for Economic Cooperation and Development, Paris, France

(21) Sanders, H.J. (1975) *Chem. Eng. News* **53**(13), 18

Appendix A

TYPICAL CONTENTS OF A QUALITY MANUAL FOR TESTING LABORATORIES

The following recommendations are adapted from a report by the International Laboratory Accreditation Conference, titled "Typical Contents of a Quality Manual for Testing Laboratories." This report was produced in part by ILAC Task Force D in October 1984 and is intended to assist a laboratory in setting forth, in a systematic way, the measures it will employ to implement its quality assurance system. Although the reader may find these recommendations unnecessarily detailed, reference to them will help to avoid overlooking any important topics in the development of the program and the associated manual. See Chapter 1, reference (*13*).

1. Table of Contents

2. Quality Policy

2.1 Objective

State the goals to be achieved by implementation of the quality assurance system. These goals are often found in a management policy statement on the subject.

2.2 Resources Employed

List the resources allocated to the program, including allotted time, personnel assignments, and any space or special equipment requirements.

2.3 Quality Assurance Management

Identify staff members responsible for developing the program and the manual, and then handling the continuing operation of the program.

3. Description of the Quality Assurance Manual

3.1 Terminology

Define any unavoidable unusual terms that are used in the quality assurance manual.

3.2 Scope

Include a brief statement of the overall quality assurance plan through which the objective (2.1) is to be achieved.

3.3 Fields of Testing Activity

Identify the specific areas of laboratory testing that are covered by the manual (e.g., heavy metals analysis, sterility, pesticide residues).

3.4 Management of the Quality Assurance Manual

Indicate the individual who is responsible for keeping the manual current and seeing to the proper distribution of copies. Describe the procedure for identifying any errors found in the manual or making suggestions for changes.

4. Description of the Laboratory

4.1 Identification

The name and address of the laboratory, of course, should be stated, and a description of the corporate affiliations or any other information required for full identity.

4.2 Fields of Activity

Describe any features of the laboratory or its operations necessary to convey a true picture of the organization, such as the location and size of branch laboratories, types of services offered, and major fields of activity.

4.3 Organizational Structure

Include an organizational chart showing lines of authority and allocation of functions, including quality assurance responsibilities.

4.4 Responsibility for the Quality Assurance System

Describe the lines of responsibility for developing and maintaining the program, including the relationship of the QA staff to other laboratory staff.

4.5 Technical Management Personnel

Identify those having technical management authority in the areas covered by the QA manual, and the lines of authority and communication between them and those responsible for the QA system (4.4).

4.6 Documentation of Employee Responsibility

Refer to written instructions and information that have been given to staff members to ensure that each employee is aware of the extent of his or her area of responsibility.

4.7 Deputy Assignments

Reference or describe alternate assignments of responsibility for exercising management and QA functions when the regularly assigned staff is absent.

4.8 Minimizing Improper Influence

Reference or state any management policies that are designed to protect the quality of laboratory tests from any improper influence that might adversely impact upon actions of the laboratory's personnel.

4.9 Proprietary Rights and Confidential Information

Indicate measures that the laboratory employs to protect proprietary and confidential information.

5. Staff

5.1 Job Descriptions

Provide or reference job or position descriptions for the senior technical staff members.

5.2 Personnel Records

Indicate how records covering the education and technical experience of the laboratory staff and those involved in the QA program are maintained. These records should include all special training for performing specific tasks or in the use of specific pieces of analytical equipment, along with normal education credentials.

5.3 Supervision of Personnel

For each technical operating unit, provide information on the number of supervisory and nonsupervisory personnel and the measures used to ensure the adequacy of supervision.

5.4 Other Measures

List or describe other activities that are designed to maintain the quality of the laboratory staff, for example, the recruitment policy, in-house training, and incentive programs.

6. Testing and Measuring Equipment

6.1 Inventory

Provide a list of all major items of test equipment and measuring instruments required to perform testing. For each major item, include the following: manufacturer, model and serial numbers, date received and date placed in service, location in the laboratory, and any maintenance details.

6.2 Identification of Equipment Subject to Calibration

Describe the criteria for identifying equipment that is subject to calibration, the procedure for indicating the date of the last calibration and the due date of the next calibration, and the alert system that flags when a recalibration is due.

6.3 Maintenance and Misuse

Describe, or reference, the periodic maintenance procedures used for analytical test equipment and include a copy of instructions to the staff on how to report malfunctioning or misused equipment.

6.4 Calibration and Checking

6.4.1 Calibration Prior to Use

Document the fact that measuring and testing equipment that is used in conducting equipment testing is calibrated prior to being placed in service.

6.4.2 Calibration Programs

Describe the overall program of calibration, identifying any outside sources used and the traceability of procedures and standards to acceptable national or international standards of measurement.

6.4.3 Restricted Use of Reference Standards

Include the directives intended to ensure that reference standards of measurement are used only for calibration purposes.

6.4.4 Checking of In-Service Testing Equipment

Include the directives that specify the frequency and conditions under which in-service testing equipment is to be checked.

6.5 Purchase and Acceptance of Equipment and Supplies

Outline the precautions to be taken when purchasing and checking equipment and equipment consumables, for example, purchase-order format, acceptance criteria, and documentation requirements.

7. Environment

Present a brief description of how the required environmental conditions in the testing areas are to be achieved, and a description of the building facilities, their location, and construction features. Include procedures for the continuous monitoring of environmental conditions, where that is required. Describe controlled-access areas and procedures for enforcement. State good housekeeping requirements. Describe measures that are to be taken to protect equipment from the effects of corrosion and other deteriorating atmospheric conditions.

8. Test Methods and Procedures

8.1 Index of Testing Documents

List all of the standards, instructions, equipment operating manuals, and reference data needed to perform the specific tests required by the manual, and show the location of these items in the laboratory.

8.2 Use of Nonstandard Test Methods

Present a description of any nonstandard test method used.

8.3 Selection of Test Methods and Testing Sequence

Describe procedures that relate to the selection of specific tests, or series of tests, covered by the manual.

9. Updating and Control of Documents Affecting Quality

Describe the procedure employed to ensure that all instructions, standards, operating manuals, and reference data used in performing tests covered by the manual are current, and designate locations in the laboratory where staff has ready access to these items.

10. Handling of Samples and Other Items to Be Tested

10.1 Receipt and Disposal

Describe the procedures for receiving, identifying, storing, and disposing of samples.

10.2 Protection

Describe proper laboratory handling of samples to preclude their exposure to the injurious effects of temperature, contamination or corrosion, and other damage.

10.3 Security

Describe practices required to maintain sample security during all stages of handling and storage.

11. Verification of Results

Describe the techniques used to check and verify calculations and data transfers, including those performed or handled by computers.

12. Test Reports

Describe, with examples, proper report formats and methods of data presentation. Also include policy and procedures concerning the correction of, or addition to, reports already issued.

13. Diagnostic and Corrective Actions

13.1 Effectiveness Reports and Corrective Actions

Describe procedures or means that the laboratory uses to secure reports on the quality and effectiveness of its reports, and, when necessary, take corrective actions.

13.2 Proficiency and Interlaboratory Comparison Testing

Identify interlaboratory check sample programs and reference the most recent series of test results obtained.

13.3 Use of Reference Material

Describe the laboratory's use of reference materials, their source, testing, and status.

13.4 Technical Complaints

Describe the method by which the laboratory responds to technical complaints.

13.5 Quality Assurance System Audits

Describe the frequency of periodic audits of the entire quality assurance system, designate who is to conduct these audits, and indicate to whom the report should be sent.

14. Records

14.1 Maintenance of Records

Describe how the laboratory maintains records pertaining to original observations, data and calculations, calibration and equipment maintenance records, and final test reports, indicating where records are held and for how long.

14.2 Confidentiality and Security

Include the management directive and policy with respect to the confidentiality of test reports and other records, and describe security procedures for ensuring confidentiality.

14.3 Historical File of Test Methods

Describe the procedure for keeping a file of test methods.

15. Subcontracting

15.1 External Equipment

Identify those items of equipment that are only available through agreements or contracts with other organizations, and the name and address of those locations. Include information that verifies that the equipment is suitable for laboratory requirements.

15.2 External Facilities

Describe the procedure used to ensure that, in those cases where test work is performed at another site, the standards, responsibilities, and obligations of the primary laboratory are fully met.

Appendix B

FORMS USED BY U.S. FEDERAL AGENCIES

Government forms are continuously updated, revised, superseded, and discontinued. These are examples of forms that have been used by the various agencies.

Form No.	Title
1	Form FDA 464 (11/85): Collection Report
2	DEA Form 7 (Apr. 1990): Report of Drug Property Collected, Purchased, or Seized
3	FD 415a (2/83): Sample Seal
4	DEA 307 (Jun. 1985): Evidence Accountability Record
5	Form FDA 421 (3/80): Sample Accountability Record
6	DEA Form 86 (Dec. 1985): Forensic Chemist Worksheet
7	Form FD 431 (5/84): Analyst Worksheet
8	Form FD 465 (4/78): Sample Summary
9	FD Form 1609 (9/83) Research Project Record (2 sides)
10a	Closed-Loop Corrective Action System
10b	Corrective Action Request Form
10c	Corrective Action Master Log

1. FLAG SURVEILLANCE ☐ COMPLIANCE ☐				2. TYPE SAMPLE	3. SAMPLE NUMBER	
4. CF NO.	5. DATE COLLECTED		6. PRODUCT CODE	7. RESP FIRM	8. PROGRAM ASSIGN CODE	9. HOURS
					1.	1.
10. RELATED SPL	11. PC	12. COLL NO.	13. SPLING DIST	14. COLL DIST	2.	2.
					3.	3.

15. PRODUCT NAME AND IDENTIFICATION *(Quote pertinent labeling including firm name and address)*

16. REASON FOR COLLECTION *(Indicate analysis needed & document assignment. Include CP No. and/or Assignment No. if applicable.)*

17. MFG CODES *(Labels, pkg, shipping containers)*

18 MANUFACTURER *(Name, address & ZIP)*

19. SHIPPER *(Name, address & ZIP)*

20. DEALER *(Name, address, ZIP & telephone)*

21. SIZE OF LOT FROM WHICH SAMPLED	22. EST VALUE	23. RECEIPT ISSUED			24. DATE SHIPPED & DOC REF
		FDA472	FDA484	NONE	

25. DESCRIPTION OF SAMPLE AND METHOD OF COLLECTION *(Number and size of units, etc.)*

26. HOW PREPARED

27. COLL IDENT ON PKG AND/OR LABEL

28. COLLECTOR'S IDENTIFICATION ON SEAL

29. SAMPLE DELIVERED TO	30. DATE DELIVERED	31. LAB
	32. LAB W/SPLIT SAMPLE	33. ORIG C/R & RECORDS TO

34. RCDS OBTAINED

a. INVOICE NO. AND DATE

b. SHIPPING RECORD *(B/L, F/B, Waybill, Affidavit, etc.)* NO. AND DATE

c. OTHER DOCUMENTS *(Affidavits from dealers, etc.)* SIGNER'S NAME AND DATE

35. REMARKS ***(If additional space is needed, attach Form FDA 464a, C/R Continuation Sheet)***

36. CONSUMER COMPLAINT NO. OR RECALL NO.	37. AMT. $	38. 704(d) SPL	39. COLLECTOR *(Typed name and signature)*
	CRED. CD. ☐ C ☐ V ☐ B ☐	YES ☐ NO ☐	

FORM FDA 464 (11/85) PREV ED MAY BE USED.

COLLECTION REPORT

SAMPLE RECORD COPY 1

Form 1 Form FDA 464 (11/85): Collection Report

APPENDIX B

★ U.S.GPO:1990-0-269-480/20537

Read instructions on reverse before completing.

U.S. Department of Justice
Drug Enforcement Administration

REPORT OF DRUG PROPERTY COLLECTED, PURCHASED OR SEIZED

1. HOW OBTAINED (Check) ☐ Purchase ☐ Seizure ☐ Free Sample ☐ Lab. Seizure ☐ Money Flashed ☐ Compliance Sample (Non-Criminal) ☐ Other (Specify)	2a. FILE NO.	2b. PROGRAM CODE	3. G-DEP ID
4a. WHERE OBTAINED (City, State/Country)	4b. DATE OBTAINED	5. FILE TITLE	
6a. REFERRING AGENCY (Name)	6b. REFERRAL ☐ Case No. OR ☐ Seizure No. No.	7. DATE PREPARED	8. GROUP NO.

9. Exhibit No.	10. FDIN (8 characters)	11. ALLEGED DRUGS	12. MARKS OR LABELS (Describe fully)	APPROX. GROSS QUANTITY		15. Purchase Cost
				13. Seized	14. Submitted	

16. WAS ORIGINAL CONTAINER SUBMITTED SEPARATE FROM DRUG? ☐ NO (Included above) ☐ YES (If Yes, enter exhibit no. and describe original container fully)

REMARKS:

17. SUBMITTED BY SPECIAL AGENT (Signature)	18. APPROVED BY (Signature & Title)

LABORATORY EVIDENCE RECEIPT REPORT

19. No. PACKAGES	20. RECEIVED FROM (Signature & Date)	21. TITLE
22. SEAL ☐ Broken ☐ Unbroken	23. RECEIVED BY (Signature & Date)	24. TITLE

LABORATORY ANALYSIS/COMPARISON REPORT

25. ANALYSIS SUMMARY AND REMARKS

26. Exhibit No.	27. Lab. No.	28. ACTIVE DRUG INGREDIENT (Established or Common Name)	WEIGHT PER UNIT ANALYZED			32. TOTAL NET	33. RESERVE
			29. Strength	30. Measure	31. Unit		

34. ANALYST (Signature)	35. TITLE	36. DATE COMPLETED
37. APPROVED BY	38. TITLE	39. LAB. LOCATION

DEA Form - 7 (Apr. 1990)   Previous edition dated 10/87 is OBSOLETE.   1 - Prosecution

Form 2 DEA Form 7 (Apr. 1990): Report of Drug Property Collected, Purchased, or Seized

Form 3 FD 415a (2/83): Sample Seal

Form 4 DEA 307 (Jun. 1985): Evidence Accountability Record

*U.S. GOVERNMENT PRINTING OFFICE: 1984—459-197

1. STORAGE LOCATION		2. NAME OF PRODUCT	3. SAMPLE NO.
A.	C.		
B.	D.	4. NAME AND ADDRESS OF RESPONSIBLE FIRM	☐ CR$_X$/DEA SPL ☐ SPLIT SAMPLE
5. DATE SAMPLE RECEIVED		6A. BY WHOM RECEIVED	6B. DIST/DIV	7. DATE RECORDS REC'D.

8. METHOD OF SHIPMENT	A. PERSONALLY FROM	C. SHIPPED FROM
	B. VIA *(Check one)* ☐ PP ☐ BUS ☐ FREIGHT ☐ AIR	D. B/L NO.

9. DESCRIPTION OF SHIPMENT	A. SHIPPING CONTAINERS	NUMBER	TYPE	CONDITION
	B. SAMPLE PACKAGES	NUMBER	SIZE, TYPE, ETC.	CONDITION
	C. SEAL INSCRIPTION	COPY IN FULL		CONDITION

10. SAMPLE DELIVERY				11. SAMPLE RETURNED			
DATE	AMOUNT	FROM	TO	DATE	AMOUNT	TO	FROM

12. SAMPLE DISPOSITION	A. DATE SDN	B. DATE DESTROYED	C. DESTRUCTION METHOD	D. AMOUNT DESTROYED	E. BY WHOM	F. REASON

FORM FDA 421 (3/80) (2 PART)   Continue on reverse; also record on reverse details for which space is lacking above   **SAMPLE ACCOUNTABILITY RECORD**

Form 5 Form FDA 421 (3/80): Sample Accountability Record

U.S. Department of Justice
Drug Enforcement Administration

FORENSIC CHEMIST WORKSHEET

Page of

RECEIVED		3. SEALS	4. FILE NO. / EXHIBIT NO. / LAB. NO.
1. From	2. Date	☐ Intact ☐ Broken ☐ None	

5. DESCRIPTION OF EVIDENCE

6. SUMMARY OF FINDINGS

7. EXH. NO.	8. LAB. NO.	9. ACTIVE DRUG INGREDIENT	10. QUANTITATIVE RESULTS	11. TOTAL NET	12. RESERVE

13. RESERVE EVIDENCE

14. FORENSIC CHEMIST'S SIGNATURE	15. DATE REPORTED

16. REMARKS:

EXHIBIT .... ____ ____ ____
99XX/NN/AA ...... ____ ____ ____
DIL X/N/A .... ____ ____ ____
BALLISTICS .... ____ ____ ____
FINGERPRINT ... ____ ____ ____
SPEC. ATTN. ..... ____ ____ ____
HOURS .... ____ ____ ____
HEROIN
HEROIN SIG. ..... ____ ____ ____
HER C/C/G .... ____ ____ ____
DIL C/C/G .... ____ ____ ____

DEA Form – 86
(Dec. 1985)

Previous edition dated 9/84 may be used.

☆U.S. Government Printing Office: 1989-241-707/08020

Form 6 DEA Form 86 (Dec. 1985): Forensic Chemist Worksheet

APPENDIX B

FLAG ANALYST WORKSHEET	1. PRODUCT		2. SAMPLE NUMBER
3. SEALS ☐ INTACT ☐ NONE ☐ BROKEN	4. DATE REC'D	5. RECEIVED FROM	6. DISTRICT OR LABORATORY

7. DESCRIPTION OF SAMPLE

8. NET CONTENTS	☐ NOT APPLICABLE DECLARE/UNIT_____ ☐ NOT DETERMINED AMOUNT FOUND_____ ___ UNITS EXAMINED % OF DECLARED_____	9. LABEL-ING	_____ ORIGINAL(S) SUBMITTED _____ COPIES SUBMITTED ☐ NONE

10. SUMMARY OF ANALYSIS

11. RESERVE SAMPLE

12. a. ANALYST SIGNATURE *(Broke Seal ☐)*	13. WORK-SHEET CHECK	a. BY
b.		b. DATE
c.		14. DATE REPORTED

FORM FDA 431 (5/84)   PREVIOUS EDITION IS OBSOLETE.   PAGE OF PAGES

U.S. GOVERNMENT PRINTING OFFICE: 1984 443-969

Form 7 Form FD 431 (5/84): Analyst Worksheet

1. PRODUCT		2. SAMPLE NO.

3a. SAMPLE CLASSIFICATION	b. LABORATORY CONCLUSIONS
(Insert classification code for each PAC No. represented by sample.)	

CODE	PAC NO.

c. NAME AND TITLE	d. SIGNATURE	e. DATE

4a. DISTRICT CONCLUSIONS

b. NAME AND TITLE *(District)*	c. SIGNATURE	d. DATE

FORM FD 465 (4/78)   PREVIOUS EDITION IS OBSOLETE.   SAMPLE SUMMARY

Form 8 Form FD 465 (4/78): Sample Summary

APPENDIX B

RESEARCH PROJECT RECORD

1. RESEARCH PROJECT NO.
2. LABORATORY
3. FISCAL YEAR(S)
4. PMS NO.
5. TITLE
6. NAME OF SCIENTIST(S)
6a. DATE SIGNED
7. CONTINUING RESEARCH
   ☐ YES *(If yes explain in Item 10)*
   ☐ NO
8. PROJECT PLAN
   STARTING DATE
   COMPLETION DATE
9. TIME

FISCAL YEAR	HOURS
TOTAL HOURS	

10. SUMMARY OF PROPOSED WORK

11. OBJECTIVES

	QUARTER			
	1st	2nd	3rd	4th
a.				
b.				
c.				
d.				
e.				
f.				

12. COMMENTS

13a. NAME OF SCIENCE ADVISOR *(DATE)*
   b. NAME OF SUPERVISOR *(DATE)*
   c. NAME AND DATE OF APPROVING OFFICIAL

FORM FDA 1609 (9/83)   PREVIOUS EDITION IS OBSOLETE.   PAGE   OF   PAGES

Form 9 FD Form 1609 (9/83) Research Project Record (2 sides)

INSTRUCTIONS FOR COMPLETING FORM FDA-1609

GENERAL

Prepare this form when a research project is being proposed or has been completed. The form must be typed and suitable for photo-duplication. Attach additional pages, as necessary, to continue items, 10, 11 or 12. Identify continuation pages by project title and number in the upper right-hand corner of each page, and "page   of   pages" in lower right-hand corner.

SPECIFIC ITEMS

1. PROJECT NUMBER. Enter the HFO-600 assigned project number if it has been assigned. Contact HFO-600 for project numbers for local projects.

2. LABORATORY. Name and mailing symbol.

3. *FISCAL YEAR(S). Enter the fiscal year or years you expect to work on the project. Also enter in block 8.

4. PMS NO. Enter Program Management System Number.

5. TITLE. Enter a *concise descriptive* title that contains the key words that might be used for publication.

6. SCIENTIST(S). Type name, sign, and date.

7. CONTINUING RESEARCH. If yes, explain clearly the most recent accomplishment and justify continuation.

8. Enter the starting or completion date for this research project when filing the initiation and termination of Form FDA 1609.

9. *Complete as indicated.   Enter the fiscal years shown in block 2 and the corresponding hours planned for each fiscal year.

10. SUMMARY. For PROPOSED PROJECTS, summarize the proposed work and background under the following headings:

a. Purpose. A statement that describes the problem that this research is intended to solve.

b. Significance. Relate this project to the FDA mission and to other related work.

c. Plan of Work. A description of the research to be done. This should include: (1) State-of-the-art in the problem area; (2) reasons for selecting the propose approach and rejecting others; and (3) planned steps to be taken to solve the problem. Include plans for collaborative study and publication.

d. Literature Review. List major references, with titles, that are related to the problem or to your approach.

e. Budget. List major equipment not available in the laboratory and justify each item and exclude expendable supplies.

For COMPLETED PROJECT, summarize the work accomplished and include the following information: (1) Time spent (by fiscal year); (2) cost of special equipment; (3) whether or not a collaborative study was completed; (4) resulting publications; (5) adoption by any compendial organization (AOAC, USP, etc.) or inclusion in FDA program or assignment.

11. OBJECTIVES. Indicate the target quarter for completion of specific objective (for multi-year projects show fiscal year).

12. COMMENTS. Identify each comment with the reviewer's name or initials.

13a,b Science Advisor and supervisor are to sign and date
   c. Approvals. Enter the signature, title, and date of the official approving the project, e.g., Branch Director, Laboratory Director, etc.

---

*For completed projects, use the actual years, dates, and times.

Form 9 Instructions for Completing Form FDA–1609

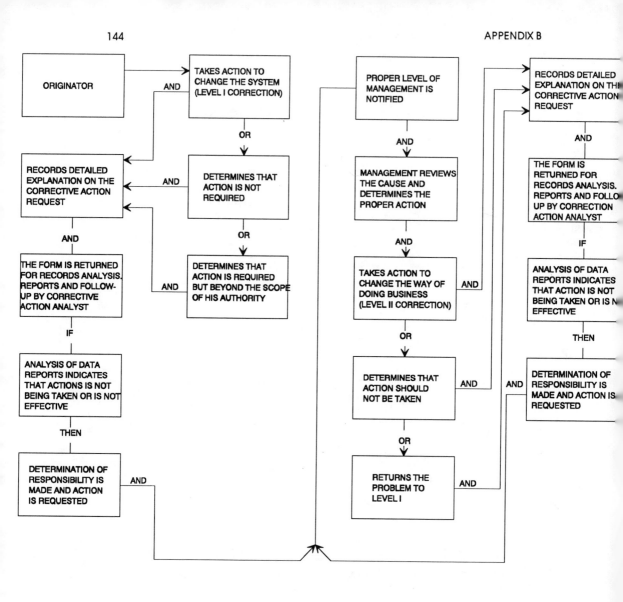

Form 10a Closed-Loop Corrective Action System

Corrective Action Request Form No. _____

NOTE: This form is part of a closed-loop corrective action system, and it is the originating organization's responsibility to follow the request through to resolution. Responsibility for a reply should be assigned to an individual rather than to an organization.

Originator _____ Date _____

Person responsible
for replying _____ Organization _____

State nature of problem below:

PROBLEM IDENFICATION

_____
_____
_____
_____

State cause of problem: _____

_____
_____
_____

Signature _____

During what phase of operation was the problem identified? _____

FROM: Name   _____
       Address   _____
       By: (Date) _____

RETURN

Information copies to: (Names and addresses)

_____
_____
_____

Form 10b Corrective Action Request Form

Corrective Action Request Form No.					
Date Submitted					
Date Received					
Submitted By					
General Problem Description					
Suspected Cause					
Assigned To					
Answer Due					
Nature of Problem Investigation					
Cause of Problem					
Corrective Action					
Corrective Action Responsibility					
Originator Notified					
Demonstration of Effectiveness					
Final Close-Out					

Form 10c Corrective Action Master Log

Appendix C

INSTRUMENT PERFORMANCE CHECKS

The following instrument performance check program has been adapted from the one provided in the *Laboratory Quality Assurance Manual of Health and Welfare, Canada*. It is offered as an example of a set of procedures that can be used to produce a continuous record of acceptable instrument performance, and to detect the first stages of deterioration in performance in an effort to provide effective preventive maintenance. Any particular laboratory may contain instruments that are not included in this list, such as liquid chromatographs or drug dissolution apparatuses, and, conversely, there are instruments included here that will not be found in every laboratory. This example, however, fully demonstrates what is an effective approach to good quality assurance documentation.

Introduction

The attached table outlines general performance checks and a schedule for conducting them. Note that this is only a general outline of test procedures, and the instrument manual must always be consulted for more detailed information about the specific instrument. Absolutely no performance checks, calibrations, or adjustments should be made without first consulting the instrument manual. Additional checks or more frequent checks may be desirable in certain situations.

Index of Instruments

- Atomic Absorption Spectrophotometers
    - Flame
    - Furnace
- Automatic Pipetting Machines
- Automatic Titrators

- Balances
  - Analytical
  - Micro
  - Top Loading
- Fluorometers
- Fluorometers, Ratio
- Fluorescence Spectrophotometers
- Gas-Liquid Chromatographs
  - General
  - Detectors
- Incubators
- Infrared Spectrophotometers
- Kjel-Foss—Fat Analyzer
- Kjel-Foss—Protein Analyzer
- Laminar Flow Hoods
- Melting Point Apparatus
- Nuclear Magnetic Resonance Spectrometers
- Ovens, General
- pH Meters
- Polarimeters
- Polarographs
- Refractometers
- Refrigerators
- Saccharimeters
- Sterilizers
  - Hot Air Ovens
  - Steam Sterilizers
- Stills, Water
- Tablet Disintegration Apparatus
  - (USP–HPB)
  - BP
- UV-Visible Spectrophotometers
- Water Baths
  - *E. coli*
  - Other Microbiological

INSTRUMENT PERFORMANCE CHECKS

Type of Instrument or Equipment	Frequency of Check	Parameters to be Checked	Standard or Reference Materials	General Procedures and/or Remarks
Atomic Absorption Spectrophotometers (Flame)	When used	1. Sensitivity 2. Detection limit	1. Standard solution of specific element to be determined. 2. Standard solution of specific element to be determined giving a response of *twice the baseline* at the highest expansion feasible.	1. Aspirate standard solutions into flame and determine absorption. Compare sensitivity against previous results. 2. Aspirate standard solution with flame 6 consecutive times. The solution that gives a minimum of twice the baseline for every aspiration represents the det. limit concentration. Compare the det. limit against previous results. a) Background corrector should be used whenever possible. b) Optimize parameters of instrument before use according to mfr's instructions. c) Record sensitivity and detection limit in instrument log book. d) If detection limit and/or sensitivity has deteriorated below values suggested in mfr's instruction manual refer to troubleshooting section of manual. If problem cannot be identified locally, a qualified service representative should be consulted. e) Report repair work (including replacement of parts) into instrument log book.
Atomic Absorption Spectrophotometers (Flame)	Quarterly	Sensitivity in frequently used AA-lamps	Same as above for sensitivity	Same as above for sensitivity.
	Yearly	General maintenance check		Maintenance work performed by qualified service representative or staff member qualified through specialized training.

Type of Instrument or Equipment	Frequency of Check	Parameters to be Checked	Standard or Reference Materials	General Procedures and/or Remarks
Atomic Absorption Spectrophotometers (Furnace)	When used	Sensitivity reproducibility	Standard solution of specific element to be determined	Inject standard solution into furnace 6 times. Compare sensitivity and reproducibility against previous results. a) Change graphite cell when instrument fails to hold preset atomizing temperature. b) Reproducibility should be within 10%. c) Refer to troubleshooting section of instruction manual for poor sensibility and/or reproducibility. Commonly, the graphite cell needs to be replaced.
	Every 40–80 injections	1. Graphite cones 2. Internal part of furnace		Clean cones and furnace according to mfr's instructions.
Automatic Pipetting Machines	Before each use	Volume of material	Graduated cylinder	
	After each day's use	Cleanliness of syringe assembly		Disassemble and clean.
	Once each week			Lubricate.
Automatic Titrators	When used	Functioning of 1. Burette 2. Response	Standard solutions	1. Eliminate air bubbles from burrette.
General Analytical Balances	When used	Level of balance zero point, cleanliness		Clean after use.
Single-Pan Analytical Balances	Quarterly	Accuracy	Reference weights	Reference weights to be calibrated and certified by National Research Council of Canada every 2 years.
	Annually or after relocation	General maintenance check and repair if required		Work to be performed by a qualified service representative or staff member qualified through specialized training.
Microbalances	When used	Accuracy	Calibrating weights	Calibrating weights have to be calibrated against reference weights on a monthly basis. Microbalance batteries to be checked for potential and replaced if necessary.
Top Loading Balances	Semi-annually	Accuracy	Reference weights	

INSTRUMENT PERFORMANCE CHECKS

Type of Instrument or Equipment	Frequency of Check	Parameters to be Checked	Standard or Reference Materials	General Procedures and/or Remarks
Fluorometers	When used	Wavelength		As per method.
Fluorometers (Ratio)	Monthly	1. Balance of light beams 2. Output waveform 3. Dynode voltage	100 ppb quinine sulfate solution Oscilloscope	
Fluorescence Spectro-photometers	Monthly	Wavelength and/or photometric accuracy and reproducibility	1 mg/100 mL quinine sulfate in $0.5N\ H_2SO_4$	Run excitation and emission spectra. Established specs. for 255 nm, 355 nm excitation peaks and 455 emission peak heights.
Gas-Liquid Chromatographs	Quarterly	Temperature of column oven	Portable indicating reference pyrometer	
	Quarterly	Gain, damping linearity and chart speeds recorder of GLC		Adjust gain and damping and check linearity and chart speeds.
	When used	Septum		Replace septum.
	Weekly (when in use)	Glass insert		Clean or replace glass insert.
		Electrometer		Balance electrometer.
	Yearly	GLC and recorder check electronics and perform general maintenance		Performed by a qualified service representative or staff member qualified through specialized training.
Gas-Liquid Chromatographs Detection Systems				
Flame Ionization (FID)	When used	Resolution, sensitivity, reproducibility, retention time, and noise level	A regularly used standard	Compare against previous runs.
Electron Capture (EC)	When used	1. Same as FID 2. Standing current	1. Same as FID	1. Same as FID.
Flame Photometric (FPD)	When used	Same as FID	Same as FID	
Coulson & Hall Electrolytic Conductivity	When used	Same as FID	Same as FID	Same as FID.

APPENDIX C

Type of Instrument or Equipment	Frequency of Check	Parameters to be Checked	Standard or Reference Materials	General Procedures and/or Remarks
Incubators	Each day	Temperature	Recording thermometer	Maintain temperature to an accuracy of ±2°C or within a given range as stipulated in methods.
	Biweekly	Temperature	Thermometer with 0.1°C divisions	As above.
Infrared Spectrophotometers	Monthly or depending on use	Resolution	Polystyrene film peaks at 3,095 resolved from one at 3,080 and one at 3,020, resolve from one at 3,015.	Scan total range of instrument from 4,000 cm^{-1} down. Typical medium resolution machine. 1.5–2.5 at 3,000 cm^{-1} 0.5–1.5 at 1,000 cm^{-1}
	As above	Wavenumber accuracy	Polystyrene film peaks at 2851.5 1601.8 1028.3	Scan total range. Typical accuracy ±3.0–5.0 cm^{-1} over 4,000–2,000 cm^{-1} ±1.5–2.5 cm^{-1} over 2,000–400 cm^{-1}
	As above	Wavenumber reproducibility	Polystyrene film	As above. Should be better than accuracy by approximately 20%.
	As above	Beam balance (100% T line)	Air	Scan total range. ±2% T.
Infrared Equipment (NACl Discs, Gas, Cells, Liquid Cells)	When used	Condition, contamination, response	L.I.B. 342 of the FDA.	Establish proper operating conditions. Check response with a standard material.
Nujol, KBr	Each new supply, and monthly	Contamination moisture	Air	
Kjel-Foss — Fat Analyzers	When used	Zero point	C_2Cl_4	Check readout; if not zero, run solvent blank and correct zero point.
	Monthly or every 500 determinations	Accuracy	Mineral oil having known and uniform specific gravity dissolved in C_2Cl_4	

INSTRUMENT PERFORMANCE CHECKS

Type of Instrument or Equipment	Frequency of Check	Parameters to be Checked	Standard or Reference Materials	General Procedures and/or Remarks
Kjel-Foss — Protein Analyzers	Once a week (When in use)	$\%N_2$	$(NH_4)_2SO_4$ solution (0.75 mg/mL)	Run aliquot of solution. Adjust meter.
	When used	$\%N_2$	$(NH_4)_2SO_4$ solution (30 mg/mL)	Run aliquot of solution. Adjust meter.
	When used	$\%N_2$	500 mg acetanilide and nicotinic acid or standard sample	
Laminar Flow Hoods	Annually	Leaks	Approved leak detector	To be checked by a qualified service representative.
	Before and after each use	Cleanliness of hood surfaces		Clean and sanitize.
	As required by service representative	HEPA filters		Replace as required.
	Monthly	Particle counts		When not in use.
	Daily	Particle counts		When sterility tests are being performed.
Melting Point Apparatuses	Monthly	Calibration of thermometer	USP melting point references	Check bath oil. Take replicate melting points of standard materials.
NMR — Spectrometers	Monthly	1. Resolution 2. Sensitivity (signal/noise) 3. Integration 4. System stability 5. Sweep calibration	1. CH_3CHO 2. Ethylbenzene (1%) TMS, $CHCl_3$ 3. Ethylbenzene (5%) $CHCl_3$ 4. 5. $CHCl_3$	1. Quartet - H - line width at $\frac{1}{2}a = 0.8$ Hz or better. 2. 15:1 or better. 3. 5 successive scans av. dev. = 4%. 4. Short term = 0.25 Hz/min. Long term = 100/Hz/16 hrs 5. On initial calibration or when problem suspected. Check for accuracy of chemical shift versus known position.
General Ovens	Quarterly	Temperature	Indicating Reference Pyrometer	
pH Meters General	Daily when in use	Accuracy and linearity	Commercially prepared buffers, USP or equivalent standard buffers.	Bracket pH value expected as closely as possible with buffer.

Type of Instrument or Equipment	Frequency of Check	Parameters to be Checked	Standard or Reference Materials	General Procedures and/or Remarks
pH Meters Microbiological	Weekly	Precision	Solution under test	Check reading of pH meter being used against another pH meter. If greater than 0.05 pH units difference report to supervisor for corrective action. Analysts should note in the logbook any change of pH electrode, or the addition of any solution to an electrode. Electrodes should be carefully washed after use, this is particularly important in microbiological uses where solutions are suspected or known to contain toxins.
Polarimeters	Monthly	Specific rotation	200 mg of quinidine sulfate (dried for 3 hrs) in 10 mL. of $0.1N$ HCl.	Do standard and blank readings Spec. rotation = $\frac{100a}{cl}$ = $+275°$ to $+287°$ Where a = corrected reading l = length polarimeters tube in decimeters c = concentration as g/100 mL solution.
Polarographs	Monthly	Calibrate voltage scale Wave shape Response Ohm's law curve Mercury drop time	$CdCl_2$ solution 1.0 mg/mL in $0.1\ M$ HCl	*Weekly:* Cleanup mercury Check Hg and N_2 supply. *Monthly:* Check Hg drop time Record $CdCl_2$ polarogram. Run Ohm's law curve. Check and clean reference electrode as necessary.
Refractometers	When used	Accuracy	Water (dist.)	Determine refractive index according to mfr's instructions.
	Quarterly	Calibration	1. Glycerol solution No. = 1.4729 @ 20°C 2. Glycerol n-octane No. = 1.3974 @ 20°C	Determine refractive index of glycerol solution and n-octane.
Refrigerators (Microbiology)	Each day	Temperature	Thermometer with 0.1°C divisions	Maintain temperature to an accuracy of ±2°C or within a given range as stipulated in methods.
Saccharimeters	When used	Calibration and linearity	Standard; sucrose solutions of 26 g/100 mL, 13 g/100 mL and 10 g/100 mL	Calibrate with 26 g/100 mL and determine linearity with other standard solution.

INSTRUMENT PERFORMANCE CHECKS

Type of Instrument or Equipment	Frequency of Check	Parameters to be Checked	Standard or Reference Materials	General Procedures and/or Remarks
Sterilizers				
a) Hot Air Ovens	Twice each day	Temperature	Thermometer with accuracy of ±2°C	
	Bimonthly	Temperature	Thermocouple	As a means of validating thermometer readings the temperature is checked at various locations by means of a recording thermocouple.
b) Steam Sterilizers	Bimonthly	Time-temperature relationships	Recording thermocouple	
	Once each week	Cleanliness inside and out		Steam supply is to be shut off, clean sterilizer inside and out.
	Once each day	Drain screens		Clean.
	Once each day	Recording chart		Check to determine if each cycle has been completed properly.
Stills (General)	All of the following checks depend on type of raw water, volume of water used, and types of tests performed.			
	Weekly	Cleanliness	Visual	No visible accumulation of scale, etc.
	Weekly	Conductivity	Conductivity meter	For systems with continuous in-line meters. Check accuracy or meter annually.
	Monthly	pH	pH meter	
	Monthly	Heavy metals (as lead)		
	Monthly	Organic Quality	As per total organic carbon determination	
Stills (Drugs)	Monthly	Current USP Specs.	As per current USP.	
Stills (Microbiology)	Monthly	Bactericidal properties	As per Sec. 4.51 of Standard Methods for the Examination of Dairy Products.	

APPENDIX C

Type of Instrument or Equipment	Frequency of Check	Parameters to be Checked	Standard or Reference Materials	General Procedures and/or Remarks
Tablet Disintegration Apparatus (USP–HPB)	Monthly	Distances Rate (strokes/min) Condition Temp. control	Check compliance USP requirements (USP XIX—pp. 650)	Temperature Control $(37° \pm 2°C$ *Rate* 28–32 cycles/minute *Stroke* (5–6 cm) *Distance* d_1 at least 2.5 cm d_2 at least 2.5 cm Check compliance with BP requirements.
UV-Visible Spectrophotometers	Twice a month	Wavelength accuracy and reproducibility	Holmium filter and Didymium filter	Check wavelength over entire UV-visible range. Maximum deviation ±1.0 nm. Run two spectra.
		Photometric accuracy and reproducibility	60 ± 0.25 mg $K_2Cr_2O_7$/liter in H_2SO_4 0.01N	Scan spectrum from 210 to 450 mm or check absorbance at following wavelength Wavelengths (nm) Absorbance (A) 235 0.747 257 0.869 313 0.293 350 0.644 Maximum deviation ±1% of full scale on all ranges; run three spectra.
			And/or	
			National Institute of Standards and Technology SRM 930 glass filter	Check against certified values. Evidence suggests this is a better check than above.
	Twice yearly	Adjustment of zero (Beam balance)	Air	

INSTRUMENT PERFORMANCE CHECKS

Type of Instrument or Equipment	Frequency of Check	Parameters to be Checked	Standard or Reference Materials	General Procedures and/or Remarks
Water Baths				
a) *E. coli* Water Bath	Twice each day when in use	Temperature	Thermometer with 0.1°C divisions contained in water bath	Do not use water bath if reading is more than ±0.05°C off the required temperature.
	Whenever water bath thermometer has been replaced and whenever water bath has been out of operation for an extensive period of time	Accuracy of water bath thermometer	Reference thermometer which has been calibrated	Water bath thermometer correction factor should be recorded and attached to water bath.
b) Other Microbiological Water Baths	As above	As above	As above	Baths should be maintained to an accuracy of ±1°C of the requirement. Water bath thermometer correction factor should be recorded and attached to water bath. Media holding water baths should have an additional thermometer placed in a water bottle which is in the bath.

Appendix D

FDA AUDIT MEASURE PROCEDURES

This Appendix sets forth the procedures used by the U.S. Food and Drug Administration to perform the following functions:
1. Audit the performance of analysts
2. Review the results of national check samples
3. Assess sample accountability
4. Audit instrument maintenance; audit chemical standards, reagents, and media; audit controlled drug substances security; and audit control of environmental factors that can affect analytical results.

A. Sample Worksheet Review

Objective

The objective of this procedure is to verify that analytical worksheets (for domestic and import samples) accurately and completely report the work done, that the worksheet supports the laboratory classification[1] and that both the worksheet and laboratory classification provide sufficient information for a decision to be made on the need for compliance action.

Frequency

Two percent of the worksheets, but no fewer than five worksheets, generated by the laboratory must be evaluated each month. This review can be conducted once a month or at several times during a month. This quality assurance evaluation does not replace normal supervisory review of all analyst worksheets.

---

[1] The laboratory classification is their conclusion whether a sample is actionable or non-actionable under the U.S. Food, Drug and Cosmetic Act.

Conduct of Measurement

Domestic and import sample worksheets selected for review should represent the laboratory's output and should include samples analyzed in mobile laboratories. The evaluation must be made using worksheets that have received a laboratory classification and are ready to leave the laboratory.

The person conducting the review should be appointed by the laboratory director. It is recommended that the review be performed by a supervisory or senior analyst and that this duty, at least initially, be rotated among several individuals. The person conducting the review should not evaluate worksheets that were previously handled by that individual.

The reviewer should direct any deficient worksheets to the director for his or her review. The director will then discuss the deficiencies with the responsible supervisory analyst. Discussions with the employee who initiated the worksheet should be held to determine what caused the defect and what corrective action is required.

Performance factors to be used are listed below:

Performance Factors

- Accuracy and Completeness of Worksheet
    1. Worksheet clearly describes the sample and its condition when received by the analyst.
    2. Worksheet fully and accurately reflects continuity and integrity of sample.
    3. Information on the worksheet is compatible with the collection report.
    4. The use of appropriate standards, reagents, and special equipment is described on the worksheet.
    5. Calculations are accurate and easy to follow and are checked; for samples classified Adverse Findings, a calculation/method check is performed. This includes checking of conversion factors, normality factors, and so forth, for accuracy of recording.
    6. If more than one person participated in the analysis, the worksheet clearly indicates who did what.
    7. Sample reserve or disposition is correctly reported.
    8. Errors are recorded by striking through the incorrect entry and inserting, initialing, and dating the new entry.
    9. Any discarded results are explained on the worksheet.
- Methodology
    10. The appropriate method is used and is referenced on the worksheet. (Analyses performed and methods used are consistent with appropriate compliance program, assignment, or instructions from collecting investigator or as directed by supervisor.)
    11. Sample is analyzed as to permit application of appropriate guidelines, if any.
    12. Deviations from reference method are explained on the worksheet.

13. Any unofficial method that was used is validated. If the classification is Adverse Findings, the validation data are recorded on or accompany the worksheet.

☐ Laboratory Classification
14. Laboratory classification is supported by information on worksheet.
15. Classification is consistent with applicable guidelines.
16. If the classification is Adverse Findings, a check analysis is performed when appropriate.
17. Data are accurately coded for entry into PODS and LMS.[2]

☐ Time Frames
18. Sample analysis time frames are met.

B. Oral Review of Worksheets

Objective

The objective of this procedure is to verify that analytical worksheets are sufficiently accurate and complete to support testimony.

Frequency

This procedure must be used at the rate of one review per analyst or technician at least once every 6 months.

Conduct of Measure

Supervisory analysts will select one sample worksheet handled by each analyst or technician whom he or she supervises. Worksheets selected for review should be at least 6 months old and should, if possible, be classified Adverse Findings.

The analyst or technician will discuss the sample analysis with a supervisor other than the one who reviewed and classified the worksheet. The analyst's supervisor should be present and should participate in the review as appropriate. Using the worksheet, the analyst should reconstruct the procedure followed and should explain the results obtained.

The "independent" supervisor who conducts the review will evaluate the discussions, using the performance factors listed below. Deficiencies should be brought to attention of the laboratory director. Results of this review should be discussed with the employee.

---

2 PODS is the Program Oriented Data System, and LMS is the Laboratory Management System, two related database programs for storing accomplishment data and for tracking assignments through the laboratory.

Performance Factors

Using the worksheet, analyst can satisfactorily explain or describe:
1. The sample, amount, and condition when received.
2. How the sample was handled to maintain its integrity.
3. The analytical procedure used including justification for the method used and the amount analyzed.
4. Why each step in the analysis was done.
5. Who did what if more than one person participated in the analysis.
6. The calculations and sample results.
7. Analyst's disposition of remaining sample.

C. Onsite Review

Objective

The objective of this review is to observe the analyst or technician in conducting typical analytical operations. Emphasis is placed on evaluating performance factors that cannot be assessed by review of written reports and records.

Frequency

Each analyst or technician must be evaluated at least once a year using this procedure.

Conduct of Measurement

The review must be conducted by the supervisor of the analyst or technician and should be conducted when the analyst is performing work that is equal to at least the average complexity that would be expected of the position and grade being evaluated. The length of the review is generally one day; however, it may be longer or shorter as necessary to observe a typical series of analytical operations. This review can also be applied to a single operation or analytical technique.

At the completion of the review the supervisor should discuss the conduct of the assignment with employee regardless of whether deficiencies are noted. Performance factors to be used for this review are listed below. Some factors are not applicable to certain types of assignments.

Performance Factors

1. Assignment is performed in a logical and orderly manner.
2. Appropriate equipment is selected and used.
3. Method as written is correctly followed. Any deviations from the method are made for good reason and are recorded on the worksheet.
4. All raw data at the time they are obtained are recorded only on the worksheet.
5. Sample and subsample containers (e.g., flasks, separators, blenders, tubes) used throughout the analyses are identified at time of use.

6. Spectra and other automatically recorded data are identified when they are recorded to ensure the correct relationship between the data and the corresponding sample or subsample.
7. All samples, reagents, media, and standards are handled and stored properly while in the custody of the analyst or technician.
8. Care is taken to ensure that all instruments and equipment used are in proper working condition before use.
9. Proper analytical technique is used. Appropriate volumetric glassware and grades of solvent, media, and reagents are used. Quantitative transfers and dilutions are performed correctly and precisely. Instruments are used properly, and appropriate controls are used to validate analytical results.
10. Proper care is exercised in avoiding sample contamination. Glassware and equipment used are clean and free from contaminants, and special cleaning procedures are used when necessary. Reference standard materials are transferred from their containers by proper technique. Appropriate precaution is used to avoid inadvertent contamination of microbiological samples, including use of gowns, gloves, masks, head covers, and booties.
11. Appropriate safety precautions are used when necessary, such as safety glasses, gloves, aprons, face masks, and fume and bacteriological hoods.

D. (Nationwide) Check Sample Program

Objective

The objective of this program is to determine the ability of laboratories to achieve compatible results when the same sample is analyzed.

Frequency

Laboratories must analyze samples when issued.

Conduct of Measure

Conduct of the program is the responsibility of Field Sciences Branch, Division of Field Operations.[3] Samples are prepared by headquarters or field laboratories, and portions are distributed to participating laboratories. After sample analysis, results are submitted to Field Sciences Branch for review. Field Sciences Branch will prepare a summary report of results.

---

3 This unit is now called Division of Field Sciences and is located in the Office of Regional Operations in Rockville, MD.

E. Sample Accountability Review

Objective

The purpose of this procedure is to ensure proper handling, storage, accountability, and disposition of samples.

Frequency

A review involving at least 10 active (sample not disposed of) and five closed (sample disposed of) Sample Accountability Records must be conducted each quarter.

Conduct of Measure

Sample Accountability Records selected for review will be randomly selected for both samples in storage and from samples that are undergoing analysis as well as for samples for which a disposition notice has been prepared.

The person conducting the review should be designated by the director or the official who is responsible for sample storage. It is recommended that this duty, at least initially, be rotated among several individuals. The person conducting the review should not have routine responsibility for sample storage.

The evaluation will require a review of records as well as physical observation of samples and the conditions under which these samples are held.

Any deficiencies that the review discloses should be brought to the attention of the director and any other official who is responsible for sample storage.

Performance factors to be used are listed below.

Performance Factors

- Active Samples
- Sample Accountability Record
    1. Information entered on this record is accurate, complete, and compatible with other documents.
    2. The record correctly reflects the handling and storage of the sample. A physical check confirms the location and storage conditions of the sample.
    3. The record is filed and maintained in an orderly manner.
- Sample Storage
    4. Sample storage is compatible with instructions.
    5. Proper sample security is maintained. Storeroom is secure. Sample seals are intact. If the seal is a temporary one, the original seal should be attached to sample package or to the worksheet. Samples in possession of analysts are appropriately secure.
    6. Sample is stored in a clean, orderly manner. Appropriate temperature and humidity levels are maintained. Sample is protected from contamination by appropriate pest control measures.
- Closed Samples

☐ Disposition
  7. Samples are disposed of in a proper manner within specified times.
  8. Sample accountability record reflects necessary information on disposition.
  9. When required, sample accountability record shows name of employee who witnessed disposition.

F. Laboratory Controls Review

Objective

Each field laboratory has a quality control program that involves:
☐ Maintenance of instruments in proper working order
☐ Adequacy of standards, reagents, media, controlled substances, and so forth
☐ Elimination of adverse environmental factors that could affect analytical results

The objective of this procedure is to periodically review the control program to ensure that it is operating satisfactorily.

Frequency

All portions of a laboratory's quality control program must be reviewed at least once every 6 months.

Conduct of Measurement

This measurement requires a review of laboratory control records. The person conducting this review should be the director or an individual whom he or she designates who is not operationally responsible for maintaining these controls.

Performance factors to be used for this review will vary as a function of the operational controls that each laboratory uses. A suggested list is given below.

Result of the laboratory control review will be reported in the following manner:
— Report the number of times adherence to *each* performance factor is measured. For most factors, the number of reviews will be the number of control records checked.
— Report the number of times that measurement of a performance factor disclosed a defect in that part of the laboratory control program.

☐ Performance Factors
☐ Instruments
  1. Newly acquired instruments are placed in operating condition and undergo performance check-outs before use in sample analysis.
  2. Control record shows conduct of appropriate maintenance and performance checks.
  3. Instruments not operating according to specifications are clearly tagged "OUT OF ORDER." Corrective action is taken to restore the item to operational conformity.

4. If a malfunction occurred, control records indicate that an assessment was made of the effect of the malfunction on sample results obtained prior to detection and what steps, if any, were taken.

☐ Standard, Reagents, Media, Controlled Substances, and So Forth
5. The strength, quality and purity of standards, solutions, reagents, media, and so forth are determined at appropriate intervals.
6. Records show tests results, date, and name of person who performed control test.
7. Only solutions, reagents, media, and so forth of sufficient quality and purity are used so as not to interfere with or lead to erroneous or misleading assay or test results.
8. Controlled drugs and other substances are handled and stored using appropriate security procedures.

☐ Environmental Controls
9. Records exist showing the maintenance of appropriate light, temperature, and humidity controls in sensitive laboratory areas.
10. Records show the proper sampling and control of contaminated air that could adversely affect sample analyses.

Appendix E

PROFICIENCY AND CHECK SAMPLE PROGRAMS

The following list of proficiency and check sample programs was compiled by the Association of Food and Drug Officials, Science and Technology Committee, and revised in June 1988:

Association of Food and Drug Officials
Proficiency and Check Sample Programs
Compiled by AFDO
Science and Technology Committee
June 1988 Revision

Part 1 — Organizations Sponsoring Proficiency and Check Sample Programs

AAB — American Association of Bioanalysts Proficiency Testing Service
205 W. Levee Street
Brownsville, TX 78520
(512) 546–5315

AACC — American Association of Cereal Chemists
3340 Pilot Knob Road
St. Paul, MN 55121

AAFCO — AAFCO Feed Check Sample Program
R.J. Noel, Chairman
Indiana State Chemist Office
Purdue University
Biochemistry Building
West Lafayette, IN 47907

AAFM — American Association of Feed Microscopists
ATTN: Janet Windsor
1118 Apple Drive
Mechanicsburg, PA 17055

AAPFCO — Association of American Plant Food Control Officials
c/o W. Penn Zentmeyer
Virginia Dept of Agriculture and Consumer Services
P.O. Box 1163
Richmond, VA 23201
(804) 786–3511

AAPCO — AAPCO Check Sample Committee
G.M. Gentry, Chairman
Florida Department of Agriculture and Consumer Services
3125 Conner Boulevard
Tallahassee, FL 32399–1650
(904) 488–9375

AOCS — American Oil Chemists Society
P.O. Box 3489
Champaign, IL 61821–0489

AORC — Association of Official Racing Chemists
Margaret Sullivan, Secretary-Treasurer
P.O. Box 19232
Portland, OR 97219
(503) 245–1631

AOSA — Association of Official Seed Analysts
c/o James N. Lair
Illinois Department of Agriculture
801 Sangamon Avenue
Springfield, IL 62706–1001

ASTM — American Society for Testing and Materials
1916 Race Sreet
Philadelphia, PA 19103

AWPB — American Wood Preservers Bureau
P.O. Box 5283
Springfield, VA 22150

CAP — College of American Pathologists
5202 Orchard Road
Skokie, IL 60077–1034
(312) 966–5700

CDC — U.S. Centers for Disease Control
Chief, Performance Evaluation Branch
Laboratory Program Office
Atlanta, GA 30333
(404) 329–3847

DOT — U.S. Department of Transportation
Transportation Systems Center, Kendall Square
Cambridge, MA 02142

EPA–1 — U.S. Environmental Protection Agency
QA Branch, EMSL
Cincinnati, OH 45268

EPA–2 — U.S. Environmental Protection Agency
QA Division, EMSL
P.O. Box 15027
Las Vegas, NV 89114
(702) 798–2100

EPA–3 — EPA Office of Enforcement
Attn: Dr. John Gillis
National Enforcement Investigation Center
Building 53, Box 25227
Denver Federal Center
Denver, CO 80225

FDA–1 — FDA, Cincinnati District
Mary Womack, Laboratory Director
1141 Central Parkway
Cincinnati, OH 45202

FDA–2 — FDA, Los Angeles District
John Stamp, Laboratory Director
1521 West Pico Boulevard
Los Angeles, CA 90015

FDA-3 — FDA, Dallas District
Darryl E. Brown, Laboratory Director
3032 Bryan Street
Dallas, TX 75204

IGLWQ — International Joint Commission
Great Lakes Water Quality Board
P.O. Box 32869
Detroit, MI 49232–2869

NCDA — North Carolina Department of Agriculture, Dept FD
Floyd Quick
P.O. Box 27647
Raleigh, NC 27611

NFDA — National Food Processors Association
1401 New York Avenue NW, Suite 400
Washington, DC 20015
(202) 639–5975

NIOSH — National Institute of Occupational Safety and Health
Chemical Reference Laboratory (PAT) (R–3)
4676 Columbia Parkway
Cincinnati, OH 45226

SSCSP — Southern State Check Sample Program
Gail Parker, Chemical Residue Laboratory
Division of Chemistry
Florida Dept of Agriculture and Consumer Services
3125 Conner Boulevard
Tallahassee, FL 32399–1650
(904) 488–9670

USDA-1 — U.S. Department of Agriculture, FSIS
P.O. Box 5080
St. Louis, MO 65315

USDA-2 — U.S. Department of Agriculture, FSIS
Chemistry Division Science Programs
Washington, DC 20250

USGS — U.S. Geological Survey
5293 Ward Road
Arvada, CO 80002
(303) 236–5345

Part 2 — Proficiency and Check Sample Programs by Sample Type

Sample Types and Compounds	Sponsoring Organization	Substrate
Aflatoxins	AOCS, NCDA	Milks, Corn Series
Alcohol—DUI	DOT	Blood
Antibiotics	USDA–2	Meat
Antibiotics	FDA	Milk
Arsenic	USDA–1	Meat and Poultry
Asbestos (No. of Fibers)	NIOSH	Filters
Bacteriology	FDA	Milk, Foods
Bacteriology—General	CDC	
Bioassay of Antibiotics	AAFCO	Feeds
Cadmium, Lead, and Zinc	NIOSH	Filters
Cations	EPA–1	Water
Cereal Components	AACC	Cereal
Clinical Parameters	AAB, CAP	
Coliform	EPA–1	Water
Cyanide	EPA–1	Water
Diesel Fuel Octane, etc.	ASTM	Diesel Oil
Drinking Water — Primary Contaminants	EPA–1	Water
Drug Screening	CDC	
Drugs	AORC, FSF	
Explosives	FSF	Various
Fatty Acids	AOCS	Fats and Oils
Fertilizers (Magruder)	APFCO	Fertilizers
Fiber Characterization	FSF	Fibers
Firearms	FSF	Bullets & Cartridge Cases

Sample Types and Compounds	Sponsoring Organization	Substrate
Flammables	FSF	Various
Fluoride	CDC	Water
Food Components	NFDA	Foods
Gasoline Octane, etc.	ASTM	Gasoline
Glass Characterization	FSF	Glass Fragments
Herbicides and Organics	EPA–2	Urine, Water
Hypothyroidism	CDC	Blood on Filter Paper
Immunohematology	CDC	Blood
Industrial Chemicals	FDA	Fish
Lead and FEP	CDC	Blood
Medicated Feeds, PFF	AAFCO	Feeds
Metals (Cations)	FDA	Food
Metals and Minerals	EPA–1, USGS	Water
Metals and Minerals	AWPB	Treated Wood
Microbiology	AAB, CAP	
Microscopic Characterization	AAFM	Feeds
Minerals	AAFCO	Feeds
Mycology	CDC	
Nutrients (N–P)	EPA–1	Water
Oil and Grease	EPA–1	Water
Organic Chemicals	USGS	Water
Organic Solvents	NIOSH	Charcoal Tube
Organics, Semi-Volatile	ASTM	Synthetic Leachate
Organochlorine Compounds	EPA–2	Adipose Tissue, Plasma
Paint Characterization	FSF	Paint Chips
Parasitology	CDC	
PCBs & Organochlorine Compounds	USDA–1	Meats and Poultry
PCBs & Organochlorine Compounds	EPA–1	Water
Pesticides	AAPCO	Formulations
Pesticides	FLOR	Acetone

Sample Types and Compounds	Sponsoring Organization	Substrate
Phosphatase	FDA	Milk
Physiological Fluids	FSF	Cloth
PKU	CDC	Blood on Filter Paper
Protein, Moisture, Fat, and Salt	USDA–1	Meat
Proximate Analyses	AAFCO	Feeds
Rabies	CDC	
Radiology — Cesium–137	EPA–2	Air Filters
Radiology — Gamma Scan	EPA–2	Water and Milk
Radiology — Iodine–131	EPA–2	Water and Milk
Radiology — Plutonium–239	EPA–2	Water
Radiology — Radium–226 and –228	EPA–2	Water
Radiology — Strontium–89 and –90	EPA–2	Water, Milk, Air Filters
Radiology — Tritium	EPA–2	Water and Urine
Radiology — Uranium–238	EPA–2	Water
Radiology — Gross alpha and beta	EPA–2	Water, Air Filters
Residues, Pesticide	USDA–1	Meats and Poultry
Residues, Pesticides & Herbicides	USEPA	Water, Drinking Water
Residues, Pesticides & Herbicides	FDA–2	Vegetable, Food
Residues, Pesticides & Herbicides	FDA–3	Vegetable, Food
Residues, Pesticides & Herbicides	SSCSP	Vegetable, Food, Standard Reference Materials
Residues, Pesticides & Herbicides	EPA–1, USGS	Water
Residues, Pesticides & Herbicides	EPA–2	Adipose Tissue, Plasma, Water, Urine
Residues, Pesticides & Herbicides	EPA–3	Pesticide Misuse Invest.
Serology and Immunology	CDC	Serum

Sample Types and Compounds	Sponsoring Organization	Substrate
Sickle Cell	CDC	Blood
Silica	NIOSH	Filters
Smalley Oil Meals	AOCS	Feeds
Somatic Cell Count	FDA	Milk
Sulfonamides	USDA–1	Meat and Poultry
Toxicology	CAP	Serum, Urine
Vitamins	AAFCO	Feeds
Vitamins	AACC	Milk, Feed, Foods
Volatile Chemicals	EPA–1	Water
Wastewater Parameters	EPA–1	Water

Appendix F
ACCREDITATION CRITERIA

The following criteria for laboratory accreditation were developed by the Association of Official Analytical Chemists under contract with the Ministry of National Health and Welfare Canada. This work was completed in a 1983 study to develop and test an analytical laboratory accreditation program for food or drug laboratories. These criteria can be used by laboratories that wish to evaluate their own potential for accreditation, or to measure the effectiveness of their quality assurance programs.

1. Applicable Documents

The Accreditation Criteria in this part were drawn, in some measure, from the following standards, guides, and programs:

1.1 ASTM–E548, Standard Practice for Generic Criteria for Use in the Evaluation of Testing and Inspection Agencies.
1.2 Z34.1, Proposed American National Standard for a Third Party Certification Program.
1.3 ISO Guide 25–1978 (Revised March 1982), Guidelines for Assessing the Technical Competence of Testing Laboratories.
1.4 15 CFR Part 7a, National Voluntary Laboratory Accreditation Procedures, Department of Commerce.
1.5 NATLAS, National Testing Laboratory Accreditation Scheme, Accreditation Standard, Part 1.
1.6 TELARC, Testing Laboratory Registration Council of New Zealand, Laboratory Accreditation in New Zealand.

2. Purpose

The purpose of this part is to establish the criteria under which the AOAC's Analytical Laboratory Accreditation Program will operate.

2.1 The Accreditation Criteria listed below will be used as the basis for judging laboratories seeking accreditation.
2.2 Although some laboratories may operate under criteria and procedures that differ from those described, such differences shall not preclude accreditation,

provided the laboratories have comparable systems that produce quality analytical results and serve the purposes of the accreditation program.

3. Definitions

3.1	General Criteria	Means information and characteristics that should be obtainable from, or found in, a reputable testing laboratory.
3.2	Specific Criteria	Means specific characteristics or requirements pertaining to such general areas as organization, human resources, material resources, and quality systems from evaluating a laboratory's capabilities, and hence its ability to comply with the requirements for accreditation.
3.3	Organization	Means a laboratory seeking accreditation.
3.4	Human Resources	Means management, supervisory, and staff personnel who are competent to perform the tests and functions in the areas for which a laboratory seeks accreditation.
3.5	Material Resources	Means the instrument, documentation, environment, structures, etc., needed to augment the elements of support and capability provided by persons.
3.6	Quality Systems	Means the total features and activities of a laboratory to produce quality work.
3.7	Quality Control	Means a planned system of activities whose design is to provide a quality product.
3.8	Quality Assurance	Means a planned system of activities whose design is to ensure that the quality control program is effective (in this program, quality assurance will be used to cover both quality control and quality assurance).
3.9	Testing Laboratory	Means a laboratory that examines, analyzes, tests, or otherwise determines the components, characteristics, or performance of materials or products.
3.10	Test Methods	Means analytical procedures or examinations to determine one or more components and characteristics of a material or product.

GENERAL AND SPECIFIC CRITERIA

4. Organization

The laboratory shall have a documented organizational structure and material resources to operate and maintain a testing capability to conduct the examinations and analyses for which it seeks accreditation, such as:

4.1 An Ownership or management structure designating all management, operating, support, and service units; identifying all key management and supervisory positions; and defining the reporting relationships that are relevant to the accreditation request.

4.2 An appropriately housed and properly maintained facility for performance of the testing work. The premises shall be protected, as required, from contamination of excessive temperature, dust, moisture, vibration, and magnetic or other types of disturbances that may adversely affect the accuracy of measurements.

4.3 Adequate storage, workspace, lighting, ventilation, and electrical and mechanical services.

4.4 Adequate measures to ensure good housekeeping and safety for personnel.

4.5 Designated analytical test areas, controlled in a manner appropriate to their designated purposes, with defined access by visitors.

5. Human Resources

The laboratory shall be staffed by personnel who are competent to manage, supervise, and perform the tests in the discipline or areas for which it seeks accreditation, such as;

5.1 A staff that has the necessary education, training, technical knowledge, and experience for their assigned functions.

5.2 A suitable proportion of supervisory to nonsupervisory staff to ensure adequate supervision.

5.3 A file for each professional, scientific, supervisory, and technical position containing a position description, education, training, and experience.

5.4 A training program assuring that new or untrained staff will be able to perform analyses and examinations to the requisite degree of precision and accuracy.

5.5 A description of such other training, operational or planned, to ensure continued competence of personnel.

6. Material Resources

The laboratory shall be equipped with instruments and other equipment to carry out analyses and examinations in the disciplines or areas for which it seeks accreditation, such as:

6.1 An inventory of the instruments and other equipment containing:
    6.1.1 name of the item,
    6.1.2 manufacturer's name,
    6.1.3 type identification and serial number,
    6.1.4 date received,
    6.1.5 location, and
    6.1.6 details of maintenance and repair.

6.2 A program that ensures performance and accuracy checks for new equipment before it is placed in service.

6.3 Instructions for operating each item of equipment, including wiring diagrams, if available.

6.4 Documented procedures for equipment that is not of established design to demonstrate that they meet required standards of accuracy.

6.5 Documented procedures for testing special equipment that must meet certain test specifications.

6.6 A plan for withdrawal of equipment from service, and return to operation when the service has been completed.

7. Quality System

The laboratory shall operate under a documented quality system appropriate to the type, range, and volume of work performed. This includes, but is not limited to, the following:

7.1 Quality Assurance Manual

The laboratory shall have a manual that covers:

7.1.1 organization and responsibilities of key individuals,

7.1.2 quality assurance objectives,

7.1.3 quality assurance procedures (system),

7.1.4 performance and systems audits, and frequency of audits,

7.1.5 corrective action procedures,

7.1.6 forms used in the system, and

7.1.7 format and frequency of quality assurance reports to management.

8. Preventive Maintenance for Equipment

The laboratory shall have a system for taking positive actions to limit failure of equipment and to ensure that equipment is calibrated and operating with the reliability required for quality results, such as:

8.1 A schedule for each piece of equipment, covering frequency, task, and services necessary to keep the item clean, calibrated, and operating, following the instructions recommended by the manufacturer or based on operating experience.

8.2 A system with specific assignments to individuals to cover preventive maintenance responsibilities.

8.3 A training program, as necessary, for individuals that need training in performance of the preventive maintenance tasks.

8.4 A record documenting completion of the maintenance and calibration tasks.

8.5 A periodically scheduled review or surveillance program, by management, to ensure that the tasks have been accomplished and recorded.

9. Quality of Supplies

The laboratory shall have a system to ensure the quality and purity of chemical reagents, solvents, gases, primary reference and secondary (house) standards, standard solutions, purified water, and volumetric glassware used in analysis, such as:

9.1 Provide for analysis of reagents, solvents, and gases, as necessary, or required by the method.

9.2 A procurement and control program for primary reference and secondary (house) standards.

9.3 Procedure for the preparation, standardization, and restandardization of standard solutions.

9.4 Specifications for volumetric glassware.

9.5 Specifications and procedures for the preparation of purified waters.

9.6 Procedures for cleaning and, as necessary, sterilizing general-use and volumetric glassware.

9.7 Procedures for maintaining stock cultures and their periodic biochemical and serological testing, when relevant.

10. Sample Handling and Record System

The laboratory shall have and maintain a sample handling and records system to suit its particular circumstances, in order to ensure the validity of test documents, such as:

10.1 Means for identifying and handling samples or items to be tested, that come into the laboratory, so there can be no question regarding identity, preservation, condition, and results of testing.

10.2 Specific operating instructions for sample collection, when sample collection is a uniquely integral part of the service performed.

10.3 A documented sample control and accountability system that ensures integrity and chain of custody from collection through delivery to the laboratory, receipt in the laboratory, and analysis and storage before and after analysis.

10.4 A record of analysis for each sample, including any calibration and instrument checks (see next section for test reports or record of analysis).

10.5 A secure system where records are held for appropriate periods of time.

10.6 A system that provides for retrievability, the presentation and the traceability of the sample source, the procedures of analysis, results, and the persons performing the analyses. (If an abbreviated report is made to the client, the report shall indicate that additional data are available on request.)

11. Test Records

The examinations and analyses carried out by the laboratory shall be covered by test records that accurately, clearly, unambiguously, and objectively present results and relevant information, such as:

11.1 Name and address of the laboratory.
11.2 A unique serial number identifying each page of the report.
11.3 Name of client, as may be necessary.
11.4 Description of the item.
11.5 Date of item receipt, examination, and analysis, as appropriate.
11.6 Description of sample preparation, prior to examination, e.g., subdividing, compositing, etc.
11.7 Reference to the examination method, any changes, modifications, and any other pertinent information, such as validation, relevant to the test procedure.
11.8 All measurements, examinations, and derived results supported by tables, graphs, sketches, charts, and photographs, as appropriate.
11.9 A statement of measurement uncertainty, where relevant.
11.10 A statement as to whether or not the sample complies with any requirement against which it was assessed.
11.11 Signature of the analyst who conducted the test, the name and title of the person accepting responsibility for the test report, and the date of issue.
11.12 If supplemental or check analyses are made by a second person, the additional test results should follow the relevant requirements of preceding paragraphs.
11.13 Particular care and attention shall be paid to the arrangement of the test report, especially with regard to presentation of the test data. Preferably, the format shall be carefully and specifically designed for each type of test, but the heading should be standardized as far as possible.

12. Test Methods and Procedures

The laboratory shall use methods and procedures dictated by the information required and by the intended use of the results, such as:

12.1 Methods and procedures required by the specifications against which the item is to be tested.
12.2 Standardized methods, if available, when specific specifications do not exist, such as the methods contained in the *Official Methods of Analysis* of the Association of Official Analytical Chemists, *Food Chemical Codex,* and the *United States Pharmacopeia.*
12.3 Documented instructions and methods prepared by the laboratory to cover special situations.
12.4 Where nonstandard test methods are used, they shall be fully documented and comparisons shall be made with standard methods or recovery experiments conducted, to demonstrate that the test methods are suitable for the purpose.
12.5 Preference should be given to: (a) methods for which reliability (accuracy and precision) has been established in collaborative or similar studies in several laboratories, (b) methods that have been recommended or adopted

by relevant international organizations, and (c) methods, when appropriate,
that are uniformly applicable to various matrices rather than those that apply only to an individual matrix.

12.6 The choice of method, equipment, and all data relevant to the work of the laboratory shall be maintained up to date and be readily available to the staff.

12.7 The choice of method and equipment, and all data transfers, should be subject to appropriate checks.

13. Validation of Performance

The laboratory shall have adequate procedures to document the validity of its analytical performance, such as:

13.1 Internal audits to check on quality control procedures.
13.2 Appropriate data review.
13.3 Participation in proficiency test programs.
13.4 Participation in interlaboratory tests.
13.5 Replicate and check analyses.
13.6 Use of standards, spiked samples, and internal check samples.

14. Deficiency Correction

The laboratory shall have adequate procedures for the identification and correction of errors, discrepancies, and deficiencies, such as:

14.1 On-the-spot or immediate corrective action to correct or repair nonconforming data or equipment.
14.2 Long-term corrective action to eliminate causes of nonconformance.

INDEX

A

AAFCO, 166
AAPCO, 167
Acceptance sampling, 66
Accreditation, *see* Laboratory accreditation
Action limits, 22
Aflatoxins, 71
Agreement on Technical Barriers to Trade, 117
American Association for Laboratory Accreditation, 118–119
American Association of Bioanalysts, 166
American Association of Blood Banks, 118
American Association of Cereal Chemists, 166
American Association of Feed Microscopists, 167
American Chemical Society, 35, 123
    Committee on Chemical Safety, 109
    specifications for distilled water, 48–49
    view on subsampling for analysis, 69
American Industrial Hygiene Association, 93, 118
American Institute of Chemical Engineers, 123
American Institute of Chemists, 123
American National Standards Institute, 119–120
American Oil Chemists Society, 167
American Public Health Association, 44
    potable tap water testing standards, 49
American Society for Quality Control, 22
American Society for Testing and Materials, 1, 49, 167
American Society of Cytology, 118
American Wood Preservers Bureau, 167
Analysts, 79–80, 161–162; *see also* Personnel considerations
    assessment of performance, 89, 95–96
    certification, registration, or licensing of chemists, 123
    examination of test samples, 92
    record keeping by, 57, 140
    training of, 34–35, 79
Analytical methods
    accreditation criteria, 179–180
    accuracy and precision, 48, 72, 77–78
    attributes/figures of merit, 72, 74
    classification by purpose/administrative propriety, 73
    comparison with official methods, 76
    control, 76–77
    performance data, 72, 78

recommendations, 84
  ruggedness testing, 76
  selection, 71–74
  validation, 70, 74–76, 180
  see also Sample analysis
Analyzed reagents, 45
Appraisal costs, 4–5
Arithmetic mean, 14, 15
Association of American Plant Food Control Officials, 167
Association of Food and Drug Officials, 93
Association of Official Analytical Chemists, 35
  accreditation criteria, 120, 174–180
  laboratory safety guidelines, 109
  Official Methods of Analysis, 47, 50, 73, 109
  volumetric glassware specifications, 50
Association of Official Racing Chemists, 167
Association of Official Seed Analysis, 167
Atomic absorption spectrophotometers, 149–150
Attribute sampling, 66
Auditors
  selection of, 97
  training, 97–98
Audits
  check sample program (nationwide), 162
  checklist, 99
  evaluation and reporting, 100
  FDA approach, 101, 158–165
  information requirements prior to, 98–99
  laboratory controls review, 164–165
  performance, 95–96, 161–162
  personnel-file appraisal, 38
  planning, 98–99
  quality assurance unit, 97
  recommendations, 102
  sample accountability review, 163–164
  sample worksheet review, 95–96, 158–160
  site visit, 99–100, 161–162
  system, 96
  worksheet review (oral), 160–161
Australian National Association of Testing Authorities, 120–121

B

Balances, 150
Bias control charts, 24
Blank measurements, 25, 45
Bulk sampling, 66

C

Calibration curve, 21, 81
Centers for Disease Control (U.S.), 168
Central limit theorem, 17–18
Central tendency measures, 14
Certified Reference Materials, 46, 76
Check samples
    analysis review, 96
    nationwide program, 162
    organizations sponsoring, 166–170
    programs by sample type, 170–172
Chemical fume hoods, 106
Chemical Hygiene Plan, 111
Chemicals
    management of supplies, 44–45
    sensitivity to environmental conditions, 80
Chlorine, in distilled water, 49
Clinical Laboratory Improvement Act of 1967, 1
Coefficient of variation, 15, 73, 92
Collaborative testing, 73, 89–90
College of American Pathologists, 93, 118, 168
Composite laboratory samples, 69
Computers and applications
    acquisition of systems and software, 61–62, 108
    costs, 62
    laboratory information management systems, 59–62
    monitoring sample accountability, 57
Confidence intervals/limits, 19, 26
Control charts
    for accreditation, 89
    control limits, 22
    construction, 22–25
    cumulative sum, 26
    interpretation, 25–26
    purpose, 74, 83
    range-based, 25, 26
    recommendations, 27
    see also Statistical applications

Control checks, 74
Correction costs, 4, 5
Cost-benefit evaluation, 4–5
Cusum charts, 26
Curve line fitting, 21

D

Data presentation, 13–14
Degrees of freedom, 18, 19, 20
Dispersion measures, 14
Distillation of water, 48
Dixon Test, 21
Drug Enforcement Administration
    check sample program, 93
    drug property collected, purchased, or seized (form), 136
    evidence accountability record, 137
    forensic chemist worksheet, 139
Duplicate analyses/measurements, 18, 23–24, 25

E

Emergency control procedures, 112–113
Employee Performance Management System, 37
Equipment
    accreditation requirements, 176–177
    installation and servicing, 41
    inventory, 42, 176
    location considerations, 105–106
    preventive maintenance, 4, 5, 41, 42–44, 81, 177
    safety, 110, 113
    sample preparation, 70
    selection and purchase, 4, 40–41, 44
    training, 36, 40, 43
    see also Instruments; Supply management
Errors, 17, 65, 69, 70–71, 77, 83

F

F-test, 19–20
Fire and explosion hazards, 107
Fluorescence spectrophotometers, 151
Fluorometers, 151
Food and Drug Administration, 73
    analyst worksheet, 57, 140
    audit measure procedures, 158–165
    Bacteriological Analytical Manual, 49
    collection report form, 55, 135
    Good Laboratory Practices, 1, 122–123

    laboratory controls review, 164–165
    onsite review, 161–162
    oral review of worksheets, 160–161
    proficiency and check sample programs, 93, 162, 168–169
    quality assurance approach, 101
    sample accountability record, 55, 138, 163–164
    sample worksheet review, 158–160
Food, Drug, and Cosmetic Act, 73
Food samples, 71
Forms
    analyst worksheet, 140
    closed-loop corrective action system, 83, 144
    collection report, 55, 135
    corrective action master log, 146
    corrective action request form, 145
    drug property collected, purchased, or seized, 136
    employee self-evaluation, 37
    evidence accountability record, 137
    forensic chemist worksheet, 139
    instrument repair control, 42
    method authorization, 77
    research project record, 142–143
    sample accountability record, 55, 138
    sample seal, 55, 137
    sample summary, 141
    Shewhart control chart, 23
Frequency density, 16
Frequency tables, 13

G

Gas-liquid chromatographs, 151
Gaussian distribution, 15
General Agreement on Tariffs and Trade, 117
Glassware, 70; *see also* Volumetric glassware
Good Laboratory Practices, 1, 122–123
    defined, 2
Grubbs Test, 21

H

Hazardous materials, safe handling of, 109, 111–112
House reference materials, 47

I

Incubators, 152
Infrared spectrophotometers, 152

Instruments
    mistakes in operation, 81
    performance checks, 8, 42, 43, 80, 130, 147–157
    standardization/calibration, 4, 42–43, 46, 81–82, 92
    vibration and noise, 106
    see also Equipment; *and specific instruments*
Internal standard method, 81
International Directory of Certified Reference Materials, 46
International Joint Commission on Great Lakes Water Quality, 169
International Laboratory Accreditation Conference, 1, 8, 118
International Organization for Standardization, 1
    accreditation criteria, 119–120
    Council Committee on Reference Materials, 46
    guides to reference materials, 47
    quality sampling plans, 67–68
    test report requirements, 57–58
International Union of Pure and Applied Chemistry
    Commission on Physicochemical Measurements and Standards, 46
Ion exchange system, 48

J
Joint Commission on the Accreditation of Healthcare Organizations, 93

K
Kjel-Foss analyzers, 152–153

L
Laboratory
    controls review, 164–165
    design considerations, 105–108
    director, responsibilities of, 2–3, 30, 31, 77, 92, 99, 104–105
    environmental conditions, 8, 43, 131
    human resources, *see* Personnel considerations
    information management systems, 59–62
    organizational structure, 175–176
    recommendations, 115
    safety, 110
Laboratory accreditation, 1
    approaches to, 118
    criteria, 119–120, 174–180
    deficiency correction, 180
    definitions, 2, 175
    good laboratory practices, 122–123
    human resources, 176
    material resources, 176–177
    national programs, 1, 120–121, 162

objectives of systems, 119, 174–175
organization of laboratory, 175–176
personnel-file appraisal, 38
preventive maintenance for equipment, 177
proficiency testing and, 93
quality system, 177
sample handling and record system, 178
standards, guides, and programs, 174
supplies quality, 178
test methods and procedures, 89–90, 179–180
test records, 178–179
validation of performance, 180
Laminar flow hoods, 153
Least-squares estimates, 21
Linear model, 21
Location parameter, 14

M

Manual, *see* Quality assurance manual
Means, 15, 16, 17
    comparison of, 19
    confidence interval for, 19
    standard error of, 17
Measures
    of central tendency, 14
    of dispersion, 14–15
Median, 14
Melting-point apparatuses, 153
Metals analysis, 48, 70–71
Methods, *see* Analytical methods
Mode, 14
Modified methods, 73
Moisture problems, 70, 80
Multiple-unit laboratory sampling, 69

N

National Bureau of Standards, *see* National Institute for Standards and Technology, 46
National Fire Protection Association, 108
National Food Processors Association, 169
National Institute for Standards and Technology, 46, 50, 76, 91
National Institute of Occupational Safety and Health, 169
    method selection criteria, 74
National Research Council, hazardous materials guidelines, 111–112
New Brunswick Laboratory Certified Reference Materials, 46

Nitric acid, 51
Nonprobability sampling, 66
North Carolina Department of Agriculture, 169
Nuclear magnetic resonance spectrometers, 153
Nuclear reference materials, 46
Null hypothesis, 19

O

Occupational Safety and Health Administration, 111
Operating characteristic curves, 66–67
Organization for Economic Cooperation and Development, 122–123
Orientation of employees, 33–34
Outliers, 15, 20–21
Ovens, 153

P

Pareto analysis, 5
Particulate materials, 70
Percentage sampling systems, 65
Permanganate test, 49
Personnel considerations
    for accreditation, 175
    employee motivation, 20
    interviews (preemployment), 33
    laboratory director, 2–3, 30, 31, 76
    nonsupervisory staff, 2, 3, 31–32
    orientation, 33–34
    performance appraisals, 31, 36–37
    personnel files, 38
    preventive maintenance monitors, 43
    qualifications and position descriptions, 32–33
    recommendations, 38
    self-evaluation, 37
    standards of performance, 32–33
    supervisors, 2–3, 31, 33, 36–37, 43, 57, 95, 96, 98
    training, 34–36
Pesticide residue analysis, 89, 92
pH meters, 153–154
Pipetting machines, 150
Planning, *see* Quality assurance planning
Plasticware, 51, 70
Polorimeters, 154
Polarographs, 154
Polychlorinated biphenyls, 78, 80
Population parameters, 17

Precision
  control charts, 23
  method, 48, 72, 77–78
Prevention costs, 4, 5
Primary standards, 44
Probability sampling, 66
Proficiency testing
  cost considerations, 4
  interlaboratory programs, 4, 31, 79, 89–90
  intralaboratory testing, 4, 31, 79, 87–88
  organizations sponsoring, 166–170
  program format, 90–92
  programs by sample type, 170–172
  recommendations, 93
  U.S. programs, 93
Purity of chemicals, 44–45

Q

Quality assurance
  defined, 2
  FDA approach, 101
  management commitment to, 1
  purpose of, 1
Quality assurance manual
  accreditation requirements, 176
  American Public Health Association, 44
  analytical methods and procedures, 8, 131
  Canadian, 43
  components, 2, 7, 127–133
  description of manual, 8, 128–129
  diagnostic and corrective actions, 132
  environmental conditions, 8, 131
  equipment testing and measuring, 8, 43, 130
  Food Safety and Inspection Service, 78, 79–80
  format, 7–9
  laboratory organization, 8, 128–129
  purpose of, 9
  quality policy, 8, 127–128
  records, 9, 133
  review and revision, 9
  sample handling, 8, 131–132
  staff, 129
  standard operating procedures, 9
  subcontracting, 9, 133
  test reports, 8, 132

updating and control of documents affecting quality, 131
verification of results, 9, 132
Quality assurance planning
committee composition and responsibilities, 6–7
cost-benefit evaluation, 4–5
elements of a program, 3
establishing a program, 2–3
objectives, 6–7, 72
Quality control
cost considerations, 4, 5
defined, 2
purpose of, 17
review of laboratory program, 164–165; *see also* Audits
see also Statistical applications
Quality system, 2, 176

R

Radioactive areas and materials, 46, 106
Random errors, 77
Range, 14, 23
control charts based on, 25, 26
defined, 15
Ranked data, 14
Rapid methods, 73
Raw data, 13
Reagents
deterioration, 45, 80
purified water, 48–49
purity specifications, 45
see also Chemicals
Receiving dock, 106
Records, 4
analytical, 56, 58–59
computerized, 59–62
equipment maintenance, 43
recommendations, 62
responsibility for keeping, 56
retention of, 59
sample collection and handling, 55
sterilization, 51
types of, 54
see also Forms
Reference Materials, 46
guides to, 47
handling, packaging, and storage, 47

Reference methods, 73
Reference standards, 46–47, 92
Refractometers, 154
Refrigerators, 154
Regression line, 21
Regulation of laboratory practices, 1
Relative standard deviation, 14, 15, 27, 92
Results validation, *see* Proficiency testing
Routine methods, 73

S

Saccharimeters, 154
Safety
    cabinets, 106
    emergency control procedures, 112–113
    equipment, 106, 110
    in facility design, 106, 110
    hazardous materials, 111–112
    laboratory, 108–109
    personal habits and safe operating practices, 113–114
    recommendations, 115
Sample analysis
    analyst training and preparation, 78–80
    computerized, 60
    control charts, 82–83
    corrective actions, 83, 179
    cost considerations, 5
    critical control points, 78
    environmental conditions, 80, 92
    error in, 69, 70
    instruments, 80–82
    interferences, 75, 79
    mistakes/blunders, 21, 48, 65, 82
    recommendations, 84
    records of, 56–59, 177–178
    worksheet review, 56–59, 95–96, 158–161
    see also Analytical methods
Sample variance, 15
Samples
    accountability, 54–58, 162–163
    authentication of, 91
    blind/double blind, 88
    collection, 55, 64–65
    comparison/split, 89
    error in preparation, 65, 69, 79

    for intralaboratory proficiency testing, 88
    population, 17
    preparation for analysis, 69, 70–71, 90–91
    proficiency and check sample program by type of, 170–172
    records of, 54–55, 57
    recommendations, 62, 84
    retention of, 59
    selection methods, 65–66
    statistical, 17, 65
Sampling
    errors, 65, 70
    plan, 64, 65–68
    procedures, 55
    recommendations, 84
    subsampling for analysis, 68–70
    variability, 65
Scatter, degree of, 14
Screening
    methods, 73
    particulate, 70
Shewart, W.A., 14
Shewhart control charts, 22–24
Significance testing, 19
Single-phase liquids, 70
Sorted data, 14
Southern State Check Sample Program, 169
Spike recovery tests, 24–25, 75
Standard addition procedures, 75, 81
Standard consensus, 73, 75
Standard deviation, 14, 15, 27
    combining estimates of, 18
    comparison of two, 19–20
    from duplicate measurements/analyses, 18, 24
    normal curve with different values, 16
    from pair of results, 18
    probability of occurrence between specified intervals of, 16
    in ruggedness testing of methods, 76
    series of mean values, 17
    within-batch, 25
Standard error, 17
Standard operating procedures, 9; *see also* Quality assurance manual components
Standard Reference Materials, 46
Standard solutions, 23, 47–48
Standards Council of Canada, 9, 121

Statistical applications
    central tendency measures, 14
    comparison of means, 19
    comparison of standard deviations, 19
    confidence interval for a mean, 19
    cost considerations, 4
    curve line fitting, 21
    data presentation, 13–14
    dispersion measures, 14–14
    F-test, 19–20
    normal distribution, 15–17
    outliers, 20–21
    recommendations, 27
    regression line, 21
    standard deviations, 16, 18, 19, 25
    t-test, 19
    tables, 20
    see also Control charts
Statistical control, 21
Statistical sampling theory, 65
Sterilizers/sterilization, 51, 155
Stills, 48, 155
Stock solutions, 48
Student's t variate, 19
Storage facilities, 106
Supply management
    accreditation criteria, 177
    bacteriological glassware, 51
    chemicals, 44–45
    cleaning glassware and plasticware, 32, 50–51
    cost considerations, 4, 5
    culture media, 49–50
    recommendations, 52
    reference standards, 46–47
    standard solutions, 47–48
    volumetric glassware, 51
    water (purified), 48–49
Survey sampling, 66
Systematic errors, 77

T

t-test, 19
Tablet disintegration apparatus, 156
Titrators, 150
Tolerance setting, 71

Trace metals analysis, 48, 51, 71
Training, 4, 32
   analyst, 34–35, 79
   audiovisual materials, 35
   auditor, 97–98
   career development, 35
   equipment maintenance, 40, 41, 43
   first aid and CPR, 113
   individual instruction, 35
   by instrument vendors, 36, 40
   methods, 35–36
   on-the-job, 34, 35
   programmed instruction, 35
   remedial, 35
   safety, 108–109
   scientific meetings, 36, 40
   short courses, 35–36
   supply management, 50
   university/college, 36
   visits to other laboratories, 36
Trend analysis, 5

U

United States Military Standards, 66–67
United States Pharmacopeia, 46, 73
U.S. Clinical Laboratories Improvement Act of 1967, 93, 118
U.S. Department of Agriculture, 93, 169
   Food Safety and Inspection Service, 78
U.S. Department of Commerce
   National Voluntary Laboratory Accreditation Program, 117, 118–119
U.S. Department of Transportation, 168
U.S. Environmental Protection Agency, 92, 112, 122, 168
U.S. Geological Survey, 170
U.S. Occupational Safety and Health Act, 107–108
U.S. Medicare Act of 1965, 93
Upper control limit, 25, 26
Upper warning limit, 25, 26
USP Reference Standards, 45, 46–47
UV-visible spectrophotometers, 156

V

Validation, *see* Proficiency testing
Variables, control chart, 23
Variability
   method, 77
   sampling, 66, 67, 69, 71
   of test results, 83; *see also* Control charts
Variance, 14, 15, 18, 26
   Ratio Tables, 20
Volatile organic constituents in samples, 71
Volumetric glassware, 48, 50–51

W

Warning limits, 22
Water baths, 157
Water, purified, 48–49, 51
Working standards, 47

Frederick M. Garfield

Frederick M. Garfield's career covers 35 years as a chemist, chief chemist, and administrator in the U.S. Food and Drug Administration, as Assistant Director of the U.S. Bureau of Narcotics and Dangerous Drugs, and as Assistant Administrator and Director of the Office of Science and Technology in the U.S. Justice Department's Drug Enforcement Administration.

Following his retirement from government, he spent 14 years with the Association of Official Analytical Chemists as a consultant and scientific coordinator. He is a Fellow of the AOAC and a past president (1973–74).